Radio Resource Management for Wireless Networks

DISCLAIMER OF WARRANTY

For a listing of recent titles in the *Artech House Mobile Communications Series,* turn to the back of this book.

Radio Resource Management
for Wireless Networks

Jens Zander
Seong-Lyun Kim

with software authors:
Magnus Almgren
Olav Queseth

Artech House
Boston • London
www.artechhouse.com

Library of Congress Cataloging-in-Publication Data
Zander, Jens.
   Radio resource management for wireless networks / Jens Zander, Seong-Lyun Kim;
   with software authors, Magnus Almgren, Olav Queseth.
      p. cm. — (Artech House mobile communications series)
   Includes bibliographical references and index.
   ISBN 1-58053-146-6 (alk. paper)
   1. Wireless communication systems.   I. Kim, Seong-Lyun.   II. Title.   III. Series.
TK5103.2   .Z35   2001   00-065068
621.3845—dc21   CIP

British Library Cataloguing in Publication Data
Zander, Jens.
   Radio resource management for wireless networks. — (Artech House mobile
   communications series)
   1. Wireless communication systems
   I. Title   II. Kim, Seong-Lyun   III. Almgren, Magnus   IV. Queseth, Olav
   621.3'845

   ISBN 1-58053-146-6

Cover design by Gary Ragaglia

International Standard Book Number: 1-58053-146-6
Library of Congress Catalog Card Number: 00-065068

10 9 8 7 6 5 4 3 2 1

Contents

Preface

The development in the field of wireless communications has been nothing short of astonishing in the past decades. We now are witnessing the transition between the mobile telephone era and the era of wireless computing. With the breakthrough advances of digital signal processing high data rate, many of the technical problems associated with the adverse and changing propagation conditions in mobile radio communication have been solved. Multimegabit data rates to portable mobile terminals are no longer science fiction, but reality. As the engineer seems to have the upper hand in this struggle against nature, very much of the development efforts are concentrated on the social struggle for scarce resources, such as the frequency spectrum and terminal battery power. The question is not only if we, as engineers, can provide wideband wireless communication everywhere, but rather, if we can afford it. These issues, the Radio Resource Management (RRM) problem are addressed in this book.

Chapter 1 provides a brief introduction to the problems in resource management and an historical outlook over the field and its relations to adjacent disciplines. In Chapter 2, a refresher of multi-user communication theory is given. Readers with a solid background in this area may omit the first section of this chapter. The last section of Chapter 2 introduces the analysis model used throughout the book. In Chapter 3, the radio resource management problem is defined in a more stringent way. Chapter 4 is probably the core chapter of the book where static resource management and cellular system design is introduced and the RUNE tool is put to work for the first time. In Chapters 5–7, various elements of resource management, such as handoffs, dynamic channel allocation, and transmitter power control are treated. Whereas the treatment in Chapters 1–7 mainly has been focused on F/TDMA type (orthogonal) waveforms, Chapters 8 and 9 deal with spread spectrum waveforms and the special problems in RRM for these systems. In particular Chapter 9, where resource management issues in CDMA systems are dealt with, is an important chapter in the book. Chapter 10 brings the focus on to RRM for data communication systems and the special characteristics and problems related to management of packet traffic. In Chapter 11, various aspects of system planning are investigated, in particular so-called "hierarchical cell structures" (HCS). The last chapter, Chapter 12, widens the scope of RRM to consider resources other than the frequency spectrum. Here also other resources are taken into account, mainly the infrastructure of fixed networks and wireless access ports. Throughout the book, examples from the application of RRM techniques, current 2G systems, as well as future 3G systems are provided.

The book is intended as a textbook for a second graduate course in wireless networks. The student and reader should be familiar with the fundamentals of radio communications, communication theory, and some queuing theory basics. Wireless networks are complicated systems, which makes the design and performance analysis inherently difficult. Several approaches are taken in the book. Classical analysis involves highly simplified models but renders easily tractable results. Slightly more elaborate models are analyzed by means of numerical analysis. The most interesting results, however, are those derived from the more realistic models for propagation and traffic conditions. Here, stochastic simulation has been the tool of choice in the field. For this purpose, RUNE, a MATLAB™-based software tool for performance analysis in wireless networks, has been included in the book. This tool was originally developed at Ericsson, but has been developed by the software authors for pedagogical use. Most of the examples in the book that require simulations have been solved using this tool. The software solutions for these examples have been provided on the CD, enabling the reader to repeat and modify the experiments in the book. In addition, a number of problems are provided for each chapter. Some of the problems are marked with an asterisk and require simulation solutions with the RUNE tools. Most of the material has been used in courses given by the authors at the Royal Institute of Technology (KTH) in Stockholm, Sweden, and in various in-house courses at Ericsson.

The authors would like to thank the software authors, Magnus Almgren at Ericsson, who is the original architect of RUNE, and Olav Queseth at KTH, who made RUNE "student friendly," for their valuable contributions and discussions around the simulation examples in the book. We would also like to acknowledge the help received in the development of the course material and early versions of the manuscript. In particular we are grateful for the contributions of Magnus Frodigh, Håkan Olofsson, Anders Furuskär, and Sverker Magnusson at Ericsson Radio Systems. Thanks to all graduate students at the Radio Communication Systems laboratory at KTH that have been instrumental in solving and designing many of the problems. Last but not least, we would like to extend our gratitude to the anonymous reviewer whose valuable comments have helped us to clarify some of the more intricate concepts in the book.

Jens Zander
Seong-Lyun Kim

1

Introduction

1.1 Historical Perspective on Radio Resource Management

The insight that energy could be transported without wires dates back to the late 19th century. As J. C. Maxwell predicted in the 1850s, Heinrich Hertz managed in 1888 to demonstrate that his 600 MHz transmitter was capable of producing a spark in his simple receiver a few meters away in his laboratory. It was, however, the Italian engineer Guillermo Marconi who first was able to make practical and commercial use of the "Hertzian wave" phenomenon in the field of communication. After the first experiments on his father's estate in 1895, his wireless devices became a commercial success, eventually making Marconi the first (but not last!) millionaire in the wireless business. From bridging a few hundred meters in his first stumbling tries, he managed to communicate across the Atlantic Ocean between Cornwall and Cape Cod, Massachusetts in 1901. In the decades to follow, wireless communications became an essential technology onboard ships. The early 1920s saw the advent of radio broadcasting, bringing wireless receivers into every home. What happened later is all well known to most of us who have seen wireless mass market successes such as TV-broadcasting, world-wide short-wave communication, satellite communications, and now in the recent decades, mobile telephony.

From an engineering perspective, the developers of wireless systems have, over time, been struggling with different fundamental design bottle-necks, or key problems, each typical to their respective phase of development. The removal of one bottleneck pushed development a quantum leap forward,

1

just to face another bottleneck. In the following we will briefly review these key problems.

Key Problem I: Path Loss—The Early Days

Wireless telegraphy became widespread in the early years of the 20th century. Receivers in those days were passive devices, mainly consisting of a simple tuned circuit (i.e., a bandpass filter tuned to the dominant frequency of the transmissions). The consequence of this design was that all the energy at the receiver output (to create a sound in an earphone or to energize and electro-magnet to pull a pen onto a strip of paper) had to be generated at the transmitter. The loss of energy over a wireless connection, the path loss, is very large, in particular over large distances. The consequence was that large and bulky transmitters, capable of radiating an enormous amount of power were dominating the scene in the early years. Needless to say, this fact severely limited the use of mobile wireless communication, except on larger ships. The advent of the electron tube amplifier (de Forest, 1915) solved this problem. Now, receivers could be equipped with amplifiers with, in principle, any amount of amplification, which could completely compensate for the path loss. This took radio into the era of radio broadcasting which spread rapidly in the 1920s. The word "radio" became synonymous with radio broadcasting to the man on the street. Within a few decades there was a radio receiver in virtually every household in the Western World. Following the early successes, sound broadcasting was followed by TV broadcasting. In the United States this occurred in the 1930s, but elsewhere, commercial success of TV had to wait for the 1950s. Wireless communication played an important role in World War II and immediately after the war, two crucial inventions revolutionized our view on wireless communication. The first was the invention of the transistor. Intended as a tube-replacement to create lightweight, low-power, and portable radios, the transistor became synonymous with the small pocket broadcast receiver in the 1950s and 1960s. In the meantime, communication engineers became increasingly aware of the next bottleneck, thermal noise.

Key Problem II: Thermal Noise

The unavoidable thermal noise, caused by the "Brownian dance" of electrons in all materials and electronic components, provided a new challenge of a different nature. No matter how much the received signals are amplified, the noise will also be amplified with them. The second key invention in the late 1940s, was the recognition of the fact that there were fundamental limits to the amount of information and the quality of reception imposed by this noise.

The trend-setting work of Claude E. Shannon, as manifested in his "A mathematical theory on communication," published in 1949, foreboded the advent of digital communications. Although not very practical at the time, the advent of the Large Scale Integrated (LSI) circuits and Digital Signal Processing (DSP) devices in the 1970s and 1980s have made it possible to push the performance of today's wireless communication systems very close to Shannon's limits. The most remarkable achievements made possible by this new way of thinking are probably the communication with satellites and deep space probes, as well as digital mobile telephones. The latter manage to maintain acceptable voice quality in the most adverse environments, such as in moving cars or even indoors.

With these new techniques, engineers have been quite successful in pushing the performance close to the constraints manifested in the laws of physics and the thermal noise. In the 1930s another fundamental problem became evident, the limited radio spectrum.

Key Problem III: The Limited Spectrum

Although this is touched upon by Shannon (bandlimited channels), it is clear that this is not entirely a technical problem. Since there is only one "ether," it is obvious that extensive and concurrent use of the same natural resource will inevitably lead to conflicts, in this case (unwanted) interference between different users. This is clear to anyone who has tried to receive a radio program on the medium wave (AM) band in the night. Hundreds of radio stations compete for the attention of the listener and in most cases the mutual interference is devastating. Since it would be possible to properly receive most of these stations if they were "alone," we see that the problem is something outside the "struggle" against nature discussed above. Rather, this problem, as with all resource sharing problems, has a social dimension. This was recognized already in the early years of radio, when the sharing of the frequency spectrum was given an administrative solution.

The International Telecommunication Union (ITU) was formed just after World War II, mainly to deal with these problems. A concept that has been in use for spectrum resource sharing since the advent of radio communication has been frequency multiplexing. The available spectrum is split into frequency bands and since early modulation schemes produced narrow-band signals, this was an excellent way to separate different users of the spectrum and to avoid unintended interference. Within the framework of the ITU, the countries of the world have taken it on to closely regulate the use of the frequency spectrum. The spectrum hierarchy starts at the ITU level where the frequency spectrum from 10 kHz to 200 GHz is split down into almost 100 bands or allocations. These in turn are assigned to "services" in a document called the *Radio Regulations* (RR). Among these services are found fixed, mobile, broadcasting, radar, amateur, and other similar uses of the radio spectrum. It may be noted that frequency allocations are not assigned to any country or any user at this level. This is normally done by the frequency management authorities in the ITU member countries. Frequencies are let to different users by various licensing arrangements. Licenses are typically issued for considerable periods of time, to match the technical/economical life span of the radio equipment used. The frequency authorities guarantee (police) that the provisions of the RR are maintained. When it comes to the lower frequency bands with long distance propagation properties, all permissions to new transmitter sites have also to be internationally coordinated. In principle this requires that for every new transmitter, the authority has to collect the consent of all other user countries that could be affected within reasonable range. Needless to say, this is an increasingly complicated matter. Rapid technical development has been a poor match to a rather slow administrative process.

Making any significant changes in the RR, for example, to allocate frequency bands to systems using new technology is a major undertaking since a consensus decision between over 170 member states of the ITU has to be reached. This is debated at a World (Administrative) Radio Conference (WARC). One reason for this is the large differences in wealth and technological development in different parts of the world. Whereas some countries require new systems with higher performance to be installed, other countries may have the view that the old technology is still viable and that investments already made in equipment, receivers, and so forth, should be protected. Major changes, if even possible to reach, may require decades of careful planning and lobbying.

The shift from (manual) land-mobile radio to (automated) mass-market telephony in the beginning of the 1980s requires a radically different solution. It is obvious that due to their sheer numbers individual users cannot be given individual frequency assignments by the frequency authorities to protect their reception quality. Instead other solutions have to be sought. In a mobile telephone system, the owner of the system, the operator, is given a license and frequency assignments. When designing his system, the operator has to organize the use of the spectrum in such a way that interference between the users of his system is kept at an acceptable level. Mobile telephone systems utilize a combination of careful planning and automatic schemes that adapt the spectrum utilization to the current user requirements. In the planning stage, typically the base station locations are carefully planned and the base stations are at pre-assigned frequencies. The choice of which base station is to be used for connecting a mobile telephone and which actual frequency channel is to be used is done automatically while the system is in operation. Examples of such first generation mobile telephone systems are the NMT system (Scandinavia, 1981), AMPS (U.S., 1984), TACS (U.K., 1984), and other systems.

Since the advent of automated mobile telephony (cellular telephony) in the early 1980s, we have seen the introduction of second generation of digital mobile telephony systems such as, for example the GSM and D-AMPS systems in the early 1990s. These, and similar competing systems, basically provide the same service as previous analog systems, but employ advanced digital signal processing to improve the range and the tolerance to interference, allowing more users into the system without compromising the speech quality.

As new wireless systems evolve to complement and replace current second generation wireless access systems, a distinct shift in design criteria can be noted. From being primarily systems for voice communication, future wireless systems will deal with data and will in particular be tailored to different multimedia applications. Due to the large impact of the Web and Web browsers as the common software platforms for various IT-applications, provisioning Internet services may be the ruling paradigm in the definition of subsequent wireless systems.

Two approaches can be seen, extending the existing 2G systems (e.g., GPRS, EDGE) where standards are already finalized and the deployment is ongoing or the third generation (and potential subsequent generations) of wireless access systems. Third generation wide-area access systems (e.g., UMTS, IMT-2000) are likely to be deployed commencing in 2002. Such systems are not *per se* much more spectrum efficient than their predecessors but are geared to provide packet and circuit switched (low level) bearer services with on-air data rates between 384 Kbps (wide-area coverage) to 2 Mbps (indoor/microcell coverage) [1]. At the same time we can see the evolution of wireless LANs, which are aiming to provide similar services but at much higher data rates (e.g., HIPERLAN II, 10 Mbps and more) in office environments. Little is yet to be said about systems beyond these (4th generation systems) except that data rates may reach 100 Mbps or more [2]. Comparing market estimates for wireless personal communication and considering recent proposals for wide band multimedia services with the existing spectrum allocations for these types of systems show that spectrum resource management remains an important topic in the near and distant future. Resource management takes on new dimensions and can no longer be restricted to be matter of spectrum utilization only. Other important components are mobile equipment power management [3], and infrastructure deployment and cost structure [4].

1.2 Fundamental Problems in Wireless Networks

In many modern systems for wireless communication, the primary goal is to provide fixed network access to a large number of mobile or stationary users. Such a communication system, where the users are dispersed over a geographical area but where their number and locations are not *a priori* known, is referred to as an area communication system or wireless network. The most common example in the early days of history is the (national) radio/ TV broadcasting systems. Here, one-way wireless connections to individual mobile or stationary listeners are provided by a collection of broadcast transmitters connected to program distribution network. A more recent example, which we will cover in somewhat more depth later in this chapter, is a mobile (cellular) telephone system. In this example, the fixed infrastructure that the mobile users are attempting to acquire services from is the Public Switched Telephone Network (PSTN). To provide the services of the PSTN, that is, to connect mobile users to fixed (or other mobile) users in the network, the network is extended by a set of radio base stations. The base stations provide the physical 2-way radio connections to the mobiles. Figure 1.1 illustrates the principles of wireless network design.[1] The network consists of a fixed network part and a wireless part. The fixed network provides connections between base stations or Radio Access Ports (RAP), which in turn provide the wireless connections to the mobiles. The RAPs are distributed over the geographical area where mobile users are provided with communication services.

This area is simply denoted as *the service area*.[2] The mobiles that are to be provided with the required service may be anywhere within the service

---

1. There are other types of wireless systems, which also could be labeled wireless networks. Of particular interest are the ad-hoc networks, where a collection of transmitter/receivers are all capable of communicating with each other in a nonpredetermined fashion. Certain Wireless Local Area Networks (WLAN) provide such examples. This will be discussed in more detail in Chapter 10.
2. In some systems where highly built-up areas are covered, or in indoor systems it would be more appropriate to talk about the service-volume.

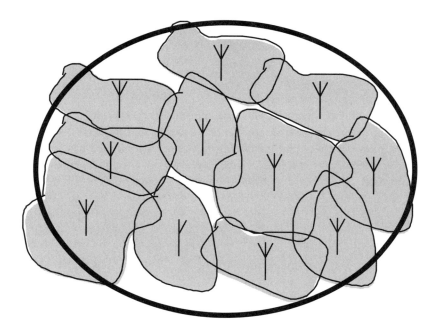

Figure 1.1 Schematic coverage map of a wireless communication system.

area and will be assigned a connection to some RAP. The assignment is usually done by the system and will be transparent to the user. The area around a RAP where the transmission conditions are favorable enough to maintain a connection of the required quality between a mobile and the RAP, is denoted the coverage area of the RAP. The transmission quality and thus the shape of these regions will, as we may expect, depend heavily on the propagation conditions and the current interference from other users in the system.

The coverage areas are usually of a highly irregular shape. They may not even be compact (contain "holes") and they usually overlap considerably. In overlapping areas a terminal may communicate with several RAPs. Unfortunately, the opposite situation may also occur where the terminal is in a region where no communication (or reception) is possible. The fraction of the service area where communication with some required quality is possible is called the coverage, or the area availability of the system. This quantity is the probability that communication is maintained at some randomly located point (with uniform distribution) in the service area. Even more interesting is the population availability, or, the probability that a randomly selected user can be provided with adequate communication service. The latter quantity can be calculated by weighting the covered areas with the (user) population density. We will discuss this more in the following sections.

In two-way communication systems, such as mobile telephone systems, links have to be established both from the RAP to the mobile (downlink or forward link) and between the mobile terminal and the RAP (uplink or reverse link). At first glance these links seem to have very similar properties, but there are some definite differences from a radio communication perspective. The propagation situation is quite different, in particular in wide-area cellular phone systems, where the RAP (base station) usually has its antennas at some elevated location, free of obstacles. The terminals, on the other hand, are usually located amidst buildings and other obstacles which create shadowing and multipath reflections. Also, the interference situation in the up and downlinks will be different, since there are many terminals and varying locations and only a few RAPs at fixed locations.

For obvious economical reasons, we would like our wireless network to provide ample coverage with as few RAPs as possible. Clearly this would not only minimize the cost of the RAP hardware and installation, but also limit the extent of the fixed wired part of the infrastructure. Coverage problems due to various propagation effects, puts a lower limit to the number of RAPs that are required. If the distance between two RAPs becomes too large, there is an obvious risk that at points between the RAP the signal-level will drop too low. Shadowing and multipath phenomena accentuate these problems. However, not quite correctly, one could say that the range of the RAPs is too small, compared to the inter-RAP distance. Such a system where this type of problem is dominant is called a *range-limited* system. Examples of range-limited systems are most mobile cellular systems in their initial stages of development when the main aim is to (quickly and at a low cost) cover the service area with adequate signal-levels and the number of subscribers is low. Other examples include the early days of broadcasting, where stations were few compared to the bandwidth available. In more mature systems, cellular and broadcasting, the number of transmitters in the system is large compared to the available bandwidth. Such systems are said to be bandwidth or interference limited.

The key problem in such a system is the proper management of the scarce resources (e.g., bandwidth) is the system to satisfy both the provider of services (the operator) and the user of these communication services. The former wants a high and efficient utilization of the system since he derives more revenues by being capable of providing services to more users. The user in turn wants a good Quality-of-Service (QoS). In mobile telephony systems, these user requirements can be expressed in terms of probabilistic measures such as the probability of being denied service when attempting to set up a connection (blocking), but also in more subjective term, for example, the sound quality of the received voice signal. For data/information services, QoS-measures, like the response time or the message delay, may be more appropriate. It will become clear later that, as in most (interesting) resource management problems, the operators aim to increase the number of users served, and the capacity of the system is in conflict with the user's desire to achieve a higher service quality. Squeezing more users into the system will inevitably cause more interference resulting either in poorer transmission quality or longer waiting times. Striking the proper balance between these aims is a delicate problem for the operator when he makes his offers to the users, particularly in a competitive situation. Efficient frequency resource management, for example, employing schemes that either avoid some of the interference or that better resist the interference between the users, can improve both the capacity as well as the service quality in the system.

Even though in this book there will be a certain emphasis on the utilization of the frequency spectrum, wireless operators and their customers have to be concerned with several other scarce resources as well. One obvious such resource is the infrastructure of networks, switches/routers, and access ports. It will become clear that a denser system with more access ports (i.e., more expensive) infrastructure, has the potential of providing more capacity and higher QoS to the users [4]. Another important resource to be managed is the power consumption in the system. Since most modern wireless networks are designed for lightweight, portable use the battery power is severely limited. Moreover, the biomedical restrictions on emitting electromagnetic fields from handheld devices also imposes limits on the transmitter power. Limitations in available energy at the portable terminal may also lead to restrictions in the complexity of the signal processing algorithms employed at the terminal. In all these cases, lower transmitter power leads to either lower transmission quality or lower radio range. Each of these effects has to be countered by the adding more access ports (i.e., a more expensive infrastructure). In a similar way to trading off power requirements and infrastructure density, it will be seen that frequency spectrum bandwidth and power can be traded off. Figure 1.2 illustrates this interdependency.

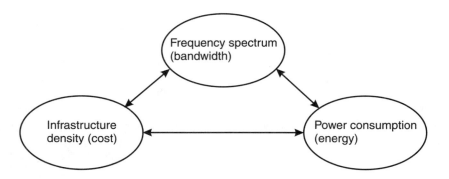

Figure 1.2 Limiting factors in wireless network design.

References

[1]   Ojanperä, T., and R. Prasad, "An Overview of Third-Generation Wireless Personal Communications: A European Perspective," *IEEE Personal Comm. Mag.*, Dec. 1998.

[2]   Flament, M., et al., "An Approach to 4th Generation Wireless Infrastructures—Scenarios and Key Research Issues," *IEEE VTC '99*, Houston, TX, May 1999.

[3]   Maguire, G. Q., et al., "Future Wireless Computing & Communication," *Nordiskt Radioseminarium, NRS-94*, Linköping, Sweden, Oct. 1994.

[4]   Zander, J., "On the Cost Structure of Future Wideband Wireless Access," *IEEE VTC '97*, Phoenix, AZ, May 1997.

2

Link Performance in Interference Channels

2.1 A Review of Multi-User Access in Mobile Communication Systems

2.1.1 Introduction and Problem Formulation

Classical communication theory deals with point-to-point links disturbed by thermal (Gaussian) noise, adverse propagation conditions, and channel variations that are difficult to predict. Real-life radio systems have to cope with additional problems. The most dominant feature of modern radio communication is that virtually no radio link or system is alone in its allocated frequency band. Other radio transmitters, near and far, constantly cause interference. Interference is, in many cases, the limiting factor to the performance of the system. With the increasing use of wireless communications, the load on the frequency spectrum has increased tremendously since the days of Marconi. A key problem area, as was already noted in Chapter 1, is how to effectively manage the frequency spectrum in order to keep the adverse effects of this interference at a minimum. Can interference be avoided, or are there efficient methods that minimize the loss in performance?

The radio transmission medium is, whether intended or not, a broadcast medium. In a wireless network, a large number of users in a geographical region attempt to communicate as illustrated by Figure 2.1. This feature is in many cases a blessing, since it enables the quick establishment of new connections between a large number of arbitrary users. In Figure 2.1, there

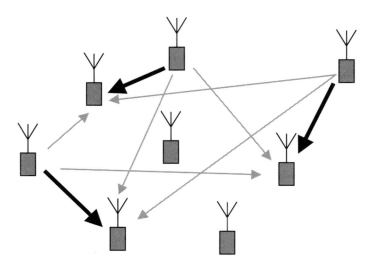

Figure 2.1 A wireless network.

are three transmitters transmitting information to three different receivers indicated by the solid black arrows. These paths are denoted the *active communication links.* However, the communication resource (the radio spectrum) is shared by all users despite the mutual interference this may cause. The transmissions of the three transmitters in Figure 2.1 give rise to interference on the unwanted cross-links as indicated by the light gray arrows. The character of the interference will depend on the waveforms and transmitter powers selected by the interfering transmitters as well as the propagation conditions on the cross-links. The impact on the performance of the active communication link will depend not only on the waveforms, the powers and the propagation conditions in the active link, but also on the detection scheme used in the receiver. In order to simplify the analysis of such a wireless network, a two-step approach is used. First the interference is characterized at the receiver by its received power and the effects of the propagation conditions. In the second step the impact of that interference on the performance of the active link is analyzed. The latter problem is the topic of this chapter.

In order to assess the impact of the interference on the performance, a given link in the network is studied. The detection scenario can be described by the multiple access channel (MAC) model in Figure 2.2. This is a straightforward extension to the M transmitter case of the models of classical communication theory. The receiver in the link of interest is facing the

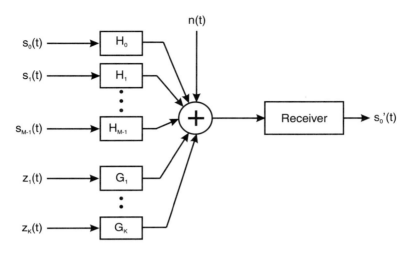

Figure 2.2 Model of signals and interference in multi-user radio system.

problem of detecting one particular transmitted signal $s_0(t)$ from some transmitter in the active link of interest. While this signal is being received, the signals $s_1(t)$, $s_2(t)$ ... $s_{M-1}(t)$ from the other $M - 1$ transmitters in the system are also on air, possibly causing interference at the receiver. As well as the known $M - 1$ transmitters in the systems, there may be additional external interference, from other (maybe distant) radio systems using the same frequency range.

Ultimately, the situation where this type of interference is deliberate and aimed at disruption of the communication links will be taken into account. For the time being, assume that there are K of these external signals and that they are denoted $z_1(t)$, $z_2(t)$ ... $z_K(t)$. Finally, as in the classical communication theory, the receiver is subject to (thermal) noise.

In many situations when the interference dominates and the noise may be neglected, the term *interference-limited* system is used. The converse, *noise-limited* systems are becoming more and more rare and are mostly found in space communications.

Methods to separate and distinguish between different users may be divided into two types: multiplexing techniques and multiple access schemes. Multiplexing describes general methods of choosing signals and combining information from different sources. In general this is done at one location, for example, a telephone switch or a microwave link carrying a large number of calls. The multiplex design problem boils down to choosing a signal constellation $s_0(t), s_1(t) \ldots s_{M-1}(t)$ such that the system achieves the required performance. Performance-criteria may differ from system to system, but in principle, they are the same as in classical communication theory. As high a data rate as possible is required at some given, low level of signal or message distortion. In the case of digital communication the bit or message error probability could be used, but also the message delay is used as a performance measure. There is, however, an important additional performance criterion that distinguishes multi-user from conventional point-to-point systems, that is, how many users are allowed simultaneously in the system at some given bandwidth. The performance of a multi-user system will depend on the selection of signals $s_0(t)$, $s_1(t)$ . . . $s_{M-1}(t)$. This problem is covered in Section 2.2.

Later in the chapter, the focus will be on the multiple access problem. The overlaying of information from a large number of users is not done in the same equipment, but rather in a distributed fashion on air (Figure 2.2). As well as the pure signal design problems, problems like synchronization and coordination of message transmissions will be encountered. Subsequent sections will deal with these problems.

2.1.2 Signal Design in Multi-User Systems

The design of communication signals in multi-user systems differs from the design of conventional point-to-point systems in several respects. To investigate these differences the discussion is started by studying optimal detection strategies for a wide class of signals in a multi-user system. However, ultimately the investigation will be confined to the class of binary, digital communication systems. Similar results can be derived also for analog schemes and for digital systems employing multilevel modulation schemes. A further assumption is that the system uses antipodal signaling, that is, the transmitters are emitting independent signals of the form

$$s_i(t) = a_i u_i(t) \qquad a_i = \pm 1 \qquad (2.1)$$

For the sake of simplicity assume that the receiver is linear and that the channel filters have a flat response (i.e., equivalent to multiplication by constant h_i). The received signal $r(t)$ may now be written as

$$r(t) = \sum_{i=0}^{M-1} a_i h_i u_i(t) + \sum_{j=1}^{K} z_j(t) = \sum_{i=0}^{M-1} a_i h_i u_i(t) + z(t) \qquad (2.2)$$

The interference vector z' consists of a number of components. The information symbol $a_0 = +1$ is assumed to be transmitted, where $z(t)$ is the sum of all the K interfering signals. The task of our selected receiver will be to determine whether the signal $s_0(t)$ or the signal $-s_0(t)$ was transmitted by transmitter zero. Equation (2.2) can now be rewritten as

$$r(t) = a_0 h_0 u_0(t) + \sum_{i=0}^{M-1} a_i h_i u_i(t) + z(t) = a_0 h_0 u_0(t) + z'(t) \quad (2.3)$$

where $z'(t)$ is the sum of the interference $z(t)$ and all (other) signal components that are independent of a_0. $z'(t)$ is thus independent of the transmitted information and can be interpreted as noise.

Using the standard vector space analogy in communication theory [1], the received signal vector r may be rewritten as the vector sum of one signal component $a_0 h_0 \boldsymbol{u}_0$ and one "noise" component z' (see Figure 2.3). If the information symbols ± 1 are equally probable, one may show [1] that the detector that minimizes the bit error probability is the one that chooses the symbol a that will maximize the probability density of the vector r. Such a maximum likelihood (ML) detector will choose $a_0 = 1$ if

$$p_r(r|a_0 = +1) > p_r(r|a_0 = -1) \quad (2.4)$$

and $a_0 = 0$ otherwise. If the constant h_0 is known, z' would constitute the only remaining stochastic component in the received vector r. The fact that $z' = r - a_0 h_0 \boldsymbol{u}_0$ enables us to rewrite the expression above by using the probability density of z' according to

$$p_r(r|a_0 = +1) = p_{z'}(r - h_0 \boldsymbol{u}_0) > p_{z'}(r + h_0 \boldsymbol{u}_0) = p_r(r|a_0 = -1) \quad (2.5)$$

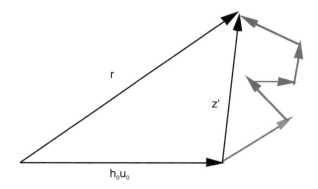

Figure 2.3 Example of vector space representation of signaler in a multi-user environment.

Now, consider the special case, where z' has a probability density function that

    i) Only depends on $|z'|$;

    ii) Is monotonically decreasing in $|z'|$;

(for instance, Gaussian noise) it can be seen that the ML receiver opts for the signal alternative that minimizes $|r - a_0 h_0 u_0|$, that is, chose the signal $a_0 h_0 u_0$ which is closest to the vector r. In this particular case, only the coordinate of r that is parallel to u_0 will be relevant to the detection process. The optimum detector for this case is the well know correlation (matched filter) detector that will determine the sign of the correlation (scalar product) $r \cdot u_0$ yielding the estimate

$$\hat{a}_0 = sgn(r \cdot u_0) \qquad (2.6)$$

In the general case, however, the conditions i) and ii) are not satisfied. In particular this will be the case when the number of interfering transmitters is small, or if a small number of interferers dominate the noise component $z'(t)$. The latter case tends to be quite common in many radio communication situations [2]. There are, however, also some interesting situations when the conditions i) and ii) are indeed satisfied. The most common situation is when $z'(t)$ consists of many signal components of roughly comparable energy. In this particular case, due to the Central Limit Theorem z' can be approximated by a zero mean Gaussian vector, thus satisfying conditions i) and ii). The receiver given by (2.6) is, in this extreme case, optimal.

Example 2.1

When detecting a BPSK-signal, the following waveform is used

$$s_0(t) = a_0 \sqrt{\frac{2E_0}{T}} \cos\left(2\pi \frac{t}{T}\right) \qquad 0 \le t < T$$

The reception is disturbed by interference that is dominated by a single PSK-modulated signal

$$s_1(t) = a_1 \sqrt{\frac{2E_1}{T}} \cos\left(2\pi \frac{t}{T} + \phi\right) \qquad 0 \le t < T$$

Other interference and thermal noise can be approximated by additive white Gaussian noise $n(t)$, with spectral density $N_0/2$. The information symbols $a_i \in \{-1, +1\}$ are independent and equally probable. The received signal $r(t)$ may be written as

$$r(t) = s_0(t) + s_1(t) + n(t)$$

a. What is the bit error probability achieved by a correlation detector according to (2.6)?

b. Is the correlation detector optimal in the ML sense?

Solution:

a. The received signal using the vector model is described in Figure 2.4 where it is assumed that the symbol $a_0 = +1$ is transmitted. The correlation receiver will only use the projection of the received vector $\boldsymbol{r}$ on $\boldsymbol{u}_0$ to make its decision. A detection error occurs if $\boldsymbol{r}$ falls in to the "wrong" half plane. If the symbol $a_0 = +1$ was transmitted, the detector will make an erroneous decision if the projection of $\boldsymbol{r}$ on $\boldsymbol{u}_0$ becomes negative.

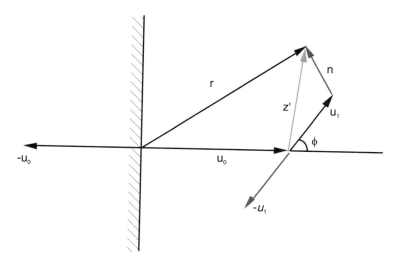

Figure 2.4 Signal constellation in example 2.1.

The projection (scalar product) is given by

$$\boldsymbol{r} \cdot \boldsymbol{u}_0 = a_1 \sqrt{E_1 E_0} \cos \phi + \sqrt{\frac{E_0 N_0}{2}} n_c$$

where n_c is a zero mean Gaussian stochastic variable with unity variance. Dividing by $\sqrt{E_0 N_0 / 2}$ the scalar product becomes negative and an erroneous decision is made if

$$n_c < -\sqrt{\frac{2E_0}{N_0}} - a_1 \sqrt{\frac{2E_1}{N_0}} \cos \phi$$

The probability of this event is given by

$$P(\text{error} \,|\, a_0 = +1; \, a_1) = Q\left(\sqrt{\frac{2E_0}{N_0}} + a_1 \sqrt{\frac{2E_1}{N_0}} \cos \phi \right)$$

Since a_1 takes the values 1 and -1 each with probability 1/2 and the situation for $a_0 = -1$ is completely symmetric the error probability may be written as

$$P_e = \frac{1}{2} Q\left(\sqrt{\frac{2E_0}{N_0}} + \sqrt{\frac{2E_1}{N_0}} \cos \phi \right) + \frac{1}{2} Q\left(\sqrt{\frac{2E_0}{N_0}} \sqrt{\frac{2E_1}{N_0}} \cos \phi \right)$$

Note here that if the interfering signal is orthogonal to $\boldsymbol{u}_0$ (i.e. $\cos \phi = 0$), it will not have any impact at all and the resulting error probability will be the same as in just the Gaussian noise with no interference present.

The error probability as a function of the signal-to-noise ratio is shown in Figure 2.5 where the signal-to-interference

$$\eta = \frac{E_0}{E_1 \cos^2 \phi} = \frac{E_0}{E_1 \theta_2}$$

is used as a parameter. θ denotes the normalized scalar product or *cross correlation* between the signals, defined as

$$\theta = \frac{\boldsymbol{u}_0 < \boldsymbol{u}_1}{|\boldsymbol{u}_0||\boldsymbol{u}_1|} = \frac{1}{\sqrt{E_0 E_1}} \int_0^T u_0(t) u_1(t) dt$$

$E_1 \theta$ can be seen as the *effective* interference energy.

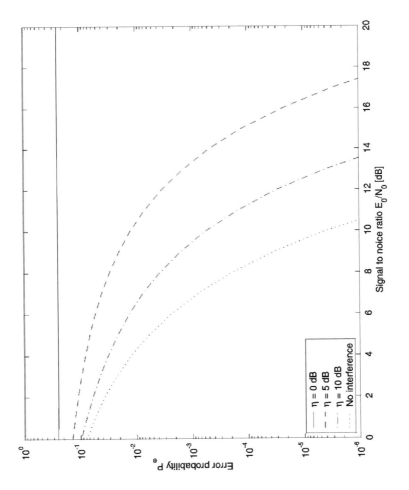

Figure 2.5 Bit error probability for correlation detector in Example 2.1 as function of the signal-to-noise ration with the signal-to-interference ratio η as parameter.

b. Let us study the interference vector

$$z' = s_1 + n = a_1 u_1 + n$$

and its probability density function

$$p_{z'}(z') = \frac{1}{2} p_n(z' - u_1) + \frac{1}{2} p_n(z' + u_1)$$

$$= C \exp(-|z' - u_1|^2/N_0) + C \exp(-|z' + u_1|^2/N_0)$$

Here, p_n is the probability density of the noise vector n. p_n has its extreme value (maximum) at the origin and satisfies the conditions i) and ii) above. $p_{z'}$ thus has two maxima, one around the vector u_1, and one around the vector $-u_1$, and cannot satisfy the two conditions. The correlation receiver is therefore not optimal for this case.

The reader is referred to [3] for a more thorough investigation of this example. Also, in the references, the cases where the signals are known up to some parameters, for example, the phase ϕ or the amplitudes are treated. Receiver designs that more or less explicitly exploit knowledge about the properties of the other interfering signals, have in the literature been labeled multi-user detectors. One has to be careful to distinguish between a true multi-user detector and an ML single user receiver. In the first case a whole set of data symbols from different transmitters are to be decoded simultaneously. This is typically the case in the base station of a mobile telephone system. In the second case only one signal is actually detected and the other data symbols are treated as unknown, but irrelevant, parameters. Example 2.1 illustrates this latter case, which is typically found in the terminal in a wireless network. The reader is referred to [4] for a more thorough treatment on advanced detection schemes.

Orthogonal Signaling

In current radio systems, most signal sets are chosen to be orthogonal, i.e. satisfying the condition

$$\mathbf{u}_j \cdot \mathbf{u}_i = 0 \qquad i \neq j,$$

corresponding to $\theta = 0$. With techniques similar to the one in the example above, one may show that if the interfering signals are orthogonal to the

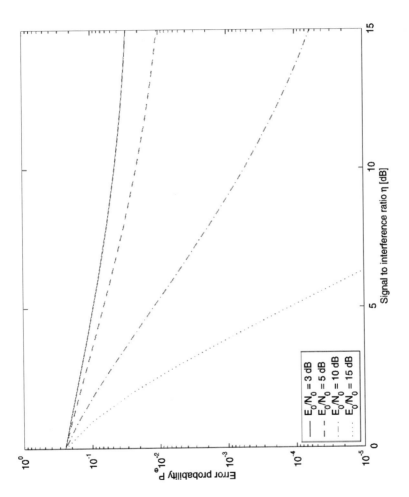

Figure 2.6 Bit error probability for correlation detector in Example 2.1 as function of the signal-to-interference ratio ration with the signal-to-interference ratio η as parameter.

wanted signal, the simple correlation receiver will be the optimum choice in the ML-sense. In fact, the effect of the interference may be completely eliminated. In Example 2.1 above, this is evident from the fact that the effective signal-to-interference ratio grows to infinity as the angle ϕ approaches 90°. The price paid for this lack of interference is that the design of orthogonal signal sets requires considerable bandwidth. It is obvious from the example that adding yet another orthogonal signal to an existing orthogonal set will require one additional dimension in the vector space. The number of dimensions, in turn, is intimately coupled to the required bandwidth. Careful studies of the properties of orthogonal signal sets [5], yield a lower bound on the required bandwidth (for any waveform). The number of orthogonal waveforms N of duration T that can exist in a band-width W is limited by

$$N \leq 2WT \tag{2.7}$$

In signal sets of the size given in (2.7) there may be signals that only differ in the carrier phase. Signal sets containing such signal pairs are, of course, not suited for noncoherent detection. If only signal sets that can be distinguished without a phase reference (i.e., possible to detect noncoherently) are considered, the relationship above becomes

$$N_{nc} \leq WT$$

It may not come as a surprise that for a constant data rate $(1/T)$ the minimal required bandwidth is directly proportional to the number of signals. Further, it may be noted that the performance (e.g., bit error probability and number of signals) does not depend on the explicit waveforms, but only on the correlation properties of the signals. Every reasonably, carefully selected set of orthogonal signals will, in principle, exhibit the same communication theoretic performance. The preference for a certain type of waveform is dictated by other reasons, typically of an implementational nature. A few of the most popular waveforms are now investigated.

Maybe the most straightforward signal set design yields the class of Time Division Multiplex (TDM) signal sets. The orthogonality condition gives

$$\mathbf{u}_j \cdot \mathbf{u}_i = c \int u_j(t) u_i(t) dt = 0$$

The simplest way to satisfy this condition is to let the integrand become zero, that is, let

$$u_j(t) = 0 \implies u_i(t) \neq 0$$
$$u_i(t) = 0 \implies u_j(t) \neq 0$$

The obvious interpretation of this is that information is transmitted in one of the signals at a time (whereas all other signals are zero). Typically the signals are chosen such that each signal $u_i(t)$ is not zero in one unique time interval—a fraction of the symbol time T. If the time intervals of the different signals (users) are not overlapping, then the orthogonality condition above is satisfied. The receiver of a certain signal may concentrate all its efforts to this particular time interval (slot) and ignore the received signal in the rest of the symbol interval. Figure 2.6b illustrates the waveform and spectrum of some time-multiplex signals.

In a similar fashion, it is possible to utilize signals that occupy disjoint frequency intervals. Using Parsevals relation

$$\mathbf{u}_j \cdot \mathbf{u}_i = c \int u_j(t)u_i(t)dt \ dt = c \int U_j(f)U_i^*(f)df \qquad (2.8)$$

where $U_i(f)$ denotes the Fourier transform of $u_i(t)$ and $U_i(f)$ its complex conjugate. It can again be noted that the signal set is orthogonal if $U_i(f)$ and $U_j(f)$ are nonoverlapping, that is,

$$U_i(f) \neq 0 \qquad U_j(f) = 0 \ \forall \ i \neq j$$

Here, different users use disjoint frequency ranges (channels) to communicate. This class of signal sets, the frequency division multiplex (FDM) is the second basic principle for designing orthogonal signal sets. Figure 2.7a illustrates these waveforms and their spectra.

A problem arising when using orthogonal signals is the impact of the channel filter on the correlation properties of the signal. To achieve interference-free communication, the signals of the signal set have to be orthogonal at the output of the channel filter. In general, this is a difficult problem since the channel affects the relative positions of the signal vectors which in turn may cause interference (cross-talk) even though the originally transmitted waveforms were orthogonal. A simple example of this is when a TDM scheme is used in a channel that has band-limited characteristics. The channel will cause time dispersion in the transmitted pulses, which

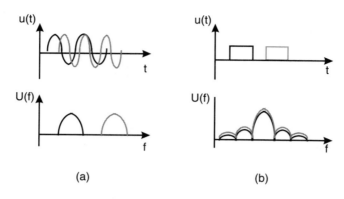

Figure 2.7 Frequency- and time-multiplex signals.

means that a message transmitted in one time-slot will partially overlap with successively transmitted messages from others. An FDM scheme is not that strongly affected by this type of channel unless the channel including the receiver can be described by a linear model. In a linear system, as in our model, no new frequency components are generated, and FDM signals remain orthogonal. However, if the channel is time varying (which is typical) or the receiver contains nonlinearities, new and overlapping frequency components may be generated (intermodulation (IM) distortion).

Using orthogonal signals is the predominant practice in most contemporary radio systems. Results from information-theory, however, indicate that there exist even larger nonorthogonal signal sets that could provide reliable communication despite the resulting interference. Unfortunately these results are not constructive, that is, no indication is given as to how a system should be designed to actually achieve these signal sets with size exceeding the orthogonal bound (2.7). This problem is further addressed in Section 2.1.4.

2.1.3 Basic Orthogonal Multiplex Schemes

Frequency Division Multiple Access (FDMA)

In radio communication history, the most popular multiplexing principle is frequency division multiplex (and frequency division multiple access, FDMA). As has been noted, FDM/FDMA means that signals that have disjoint (nonoverlapping) spectra are used. In practice, the available bandwidth is subdivided into a large number of narrow band-pass channels. If two stations choose to communicate, they (in some way) select a vacant channel. The optimum matched filter receiver consists in principle of a narrow band-pass filter selecting the appropriate signal. After this, the signal may be easily detected since the filtering process eliminates all the adjacent channel interference. All kinds of bandpass modulation techniques, analog as well as digital, with bandwidths small enough to fit into the bandpass channels may be used to convey the information (Figure 2.8).

There are several distinct advantages that have made FDMA techniques immensely popular during the entire radio communication era. The main reason is perhaps that FDMA schemes are well suited for analog circuit technology. The basic operations in FDMA transmitters and receivers are filtering and mixing of high-frequency signals for which passive and active analog circuits are ideal. Another advantage is simplicity. Although FDMA systems need to achieve a reasonable accuracy in frequency, they do not require any time or phase synchronization. The transfer of information in the separate channels occurs independently of each other, which for instance, allows the mixing of analog and digital information in the radio system.

In the early days of radio, when information rates were low compared to the available bandwidths, there was no problem in separating signals well enough in the frequency domain. The result was that the requirements on filters and absolute frequency accuracy were rather moderate. Transmitters and receivers were simple and robust devices.

With the demand for higher data rates and the ever increasing number of users, more and more signals are forced to share a limited bandwidth. These developments made increasing demands for frequency accuracy and receiver selectivity, that is, the capability of the receiver to extract the wanted signal from the multitude of signals on air. In addition, transmitters are

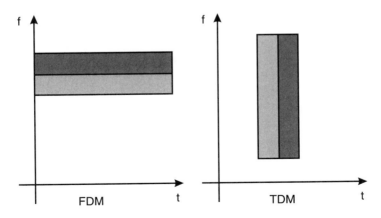

Figure 2.8 Time/Frequency diagram for TDMA and FDMA signals in a system with two active stations.

required to emit very few unwanted signal components, so-called spurious signals (lacking orthogonality), in order not to interfere with adjacent signals. In radio systems, this is a very important problem, since the dynamic range of signals at the receiver, that is, the ratio of powers between the strongest signal and the weakest signal, may be extremely large. As an example of this, consider a receiver receiving a weak signal from a distant station at the same time as it receives a signal (in an adjacent channel) but with a signal power more than 100 dB larger than the weaker signal. Extreme linearity is required in the receiver amplifiers to ensure that the output signal will still contain only the sum of these input signals, without any harmonic and/or intermodulation (IM) products appearing at frequencies other than the original signals. The design of a receiver capable of handling these situations is complex and costly.

Ever more narrowband signals introduce ever more stringent requirements on frequency accuracy in both transmitters and receivers. This causes problems mainly at high carrier frequencies (i.e. in the GHz range). The most severe drawback of FDMA technology is that the bulk of the operations in the receiver involves radio-frequency (RF) bandpass filtering, which is inherently an analog process. The production process of such devices requires either expensive high-precision analog components, or costly manual adjustments of every single device. In addition, analog RF technology is less suited for high-density VLSI implementation.

Practical examples of FDMA-based systems can be readily found in day-to-day life. In particular the analog cellular telephony systems could be mentioned. The Nordic Mobile Telephone system (NMT) was the first large-scale, commercial, fully automatic, wireless telephone system (1981). This system, and its similar systems (AMPS, TACS, and so forth) soon to follow in other countries, use a large number of narrowband channels of 25 kHz (in the United States 30 kHz) bandwidth for analog FM transmission between base stations and mobile stations. A truly full duplex voice communication link is provided by these systems and radio transmission takes place in two channels, the mobile-base (up) and base-mobile (down) channel. These duplex channels are generally separated by 10-20 MHz to allow for simultaneous transmission and reception in the mobile stations. Most analog telephone systems initially operated in the 450 MHz range, but soon expanded into the 900 MHz range.

Time Division Multiple Access (TDMA)

A multiple access scheme highly suited for digital transmission is time multiplexing (time division multiple access, TDMA). Instead of assigning only a small part of the available bandwidth to each station, all stations use the entire signal bandwidth but are confined to short, nonoverlapping time-intervals. Usually, the stations are assigned short time slots, which are repeated in a cyclic (round robin) fashion. The modulation schemes that can be used for FDMA systems can be used here as well, provided they are scaled to the larger bandwidth. It must, however, be noted that the transmitted information has to be in some time-discrete representation. The number of time slots in one cycle (or frame) is as large as the maximum number of stations that are capable of communicating simultaneously. The number of slots is equivalent to the number of channels in FDMA systems (Figure 2.8). Provided the same modulation scheme is used in both systems, the same amount of information can be transferred in the same interval in a given bandwidth in both systems.

The requirements for selectivity and frequency stability are considerably lower than for an FDM system. The task of the receiver bandpass filtering is only to eliminate out-of-band interference, and not to distinguish between different transmitters in the band used. Instead, the requirement of time accuracy, that is, synchronization, is considerable. The receivers are required to distinguish their particular time slot from the time slots of other users. This may not be an easy task, in particular if the propagation delays in the systems are comparable to the slot duration. A synchronization scheme has thus to solve two problems:

1. The classical problem of determining in which slot the wanted signal is located (this may even vary from frame to frame in some systems);
2. Transmitting/receiving accurately within the wanted slot in order to avoid overlap with other signals (lacking orthogonality).

Achieving nonoverlapping signals in all the receivers in the system is not always easy. The simplest solution is to leave some fraction at the edges (so-called guard intervals) of the time slot unused, thus allowing for small timing errors. This is analogous to leaving some portion of the spectrum between signals unused in an FDMA system. The obvious drawback of introducing guard intervals is the waste of time, which will lower the effective data rate that can be achieved in the system. When a high-efficiency is required, effective synchronization schemes are needed in order to keep the required guard intervals as short as possible. In star-shaped radio networks where many stations communicate only with a central station (e.g., mobile telephony) there is an effective solution to this problem. Here, the clock of the central station may be used as a time reference. Knowing the propagation delay to the central station would enable the peripheral stations to adjust their clocks. The peripheral stations would then transmit their messages slightly early in order to let the central controller receive the packet exactly with the proper time slot. The required remaining guard interval has now only to be in the order of the timing inaccuracy of the path delay estimates, not the whole path delay. However, as the following example illustrates, this scheme will not work if the network is not star-shaped (hierarchical).

Example 2.2

A radio system consisting of three stations in a network for tactical communication has a topology illustrated by Figure 2.9. The system uses TDMA in a half-duplex mode, that is, transmission and reception of messages cannot be simultaneous. Each station will transmit in its own time slot, and can receive messages from the other two stations in the two other time slots. The stations are using a fixed time reference and begin their transmission exactly at the beginning of a time slot. What is the minimum guard interval τ_g that is required to avoid all message overlapping. Can the starting times of message transmissions be delayed to avoid overlap?

Solution:

The speed of light can be expressed as 300 m/μs. The path delays τ_{ij} between the stations can thus be computed to be 10, 15, and 20 μs. Study the

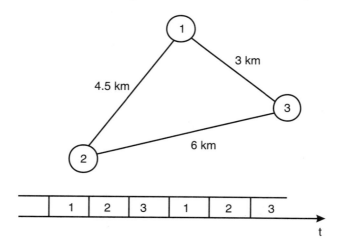

Figure 2.9 Example 2.2.

messages during reception in Figure 2.9. The most severe overlap is found in receiver 3 (time slot 3) while transmitting during the reception of a message from station 2. A message in slot 2 overlaps by 20 μs in slot 3. The guard time at the beginning (or end) of each slot must thus be at least 20 μs. In general τ_g is given by

$$\tau_g \geq \max \tau_{ij}$$

Can better results be achieved by adjusting the starting times of the transmissions? It will be seen that by delaying the starting time of transmissions by 5 μs in slot 3 the maximal overlap, and thus the guard interval can be reduce to only 15 μs.

Due to the synchronization problems, TDMA-based systems had not been common in radio communication, and time multiplexing had been mainly confined to wired transmission systems. The advent of digital signal processing and VLSI technology has radically changed this, and many TDMA-based radio systems have been developed in recent years. Maybe the most spectacular examples are found among the digital mobile telephony systems, for example, the pan-European GSM system (Global System for Mobile communication). The network structure of these systems is quite similar to the structure of their analog counterparts.

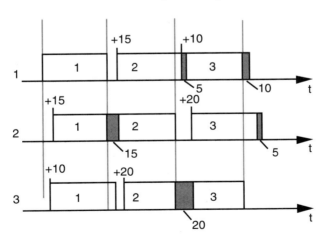

Figure 2.10 Timing diagram for reception of messages at different stations in Example 2.2. The largest message overlap is found at station 3 where two messages from station 2 and 3 overlap by 20 μs.

Regarding channel access, there are, of course, large differences. Both the D-AMPS as well as the GSM-system can be said to be FDMA/TDMA hybrids. The GSM system uses a number of approximately 200 kHz wide FDMA-channel, each of which in turn is subdivided into 8 TDMA channels for speech traffic and control information (Figure 2.11). Frames of slots of 0.577 ms each are repeated about 220 times per second. In each slot, each station transmits a burst containing 114 information bits. The gross data rate in the system is 271 Kbps. The base-mobile and mobile-base transmission occurs on separate frequency channels, even though this, at least in principle, would not be necessary. The system utilizes path delay compensation, keeping the guard interval to a relatively low value. From Figure 2.11 it can be seen that the guard interval corresponds to roughly 8 bits or about 30 μs. The fraction of wasted time is as low as approximately 5%.

2.1.4 Spread-Spectrum and Nonorthogonal Multiplexing

There is a group of multiple access techniques that are not easily classified in terms of time and frequency multiple access. These methods are often, truly or falsely, denoted as spread-spectrum or code division multiple access (CDMA) schemes. These schemes are characterized (similar to TDMA) by signals with a bandwidth much larger than 1/T. The two most popular

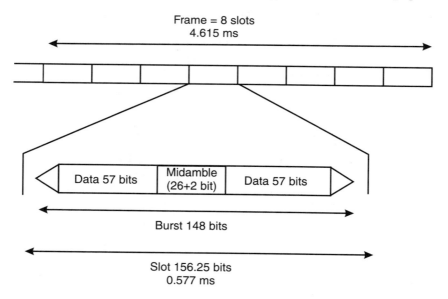

Figure 2.11 TDMA-frame structure in GSM system for mobile telephony system [6].

schemes in this class are denoted frequency hopping (FH) systems and direct sequence (DS) systems. These two techniques will now be studied briefly.

Frequency Hopping Systems

An FH system is, in principle, a combined time and frequency multiple access scheme. Similar to the traditional FDMA systems the available bandwidth is divided into a number of narrow channels. In addition, similar to TDMA, time is also divided into slots. The stations transmit narrowband signals in one of the channels during a time slot, a chip. In the subsequent time slot the station keeps transmitting, but on a new frequency channel. The station thus "hops" from frequency to frequency (Figure 2.12). The sequence of frequencies used by the transmitter is denoted as the hop sequence. All transmitters use unique but predetermined hop sequences. The (narrowband) receiver follows the same hop sequence, thus tracking the transmitter in every time slot. Since the signal in every slot is of a narrowband character, detection is done with conventional (FDMA) techniques. If the hop sequences are chosen such that no chips will overlap, it is obvious that the FH signals are orthogonal and such a system will be capable of transferring the same amount of information as an FDMA or TDMA system occupying the same bandwidth.

One usually distinguishes between fast frequency hopping (FH) systems and slow frequency hoppers (SH). In a fast hopping scheme, only one symbol (or less) is transmitted in every time slot. The hopping rate is thus equal (or larger) than the data rate. In a slow hopping system, the hopping rate is less than the data rate and several symbols, or even whole messages are transmitted in each chip. An example of the latter is GSM where the specifica-

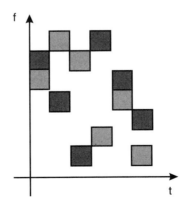

Figure 2.12 Time/frequency diagram for an FH system with two active transmitters.

tion allows for frequency hopping, transmitting one entire burst (>100 bits) on each frequency.

Unfortunately, the FH system combines the two largest drawbacks of the time and frequency multiplexing schemes. Here, both frequency selectivity (to distinguish between users) as well as very accurate synchronization are required. The latter is imperative in order to be able to track the transmitter with any success. Additional complexity is due to the rapid change of frequency required. What are the advantages then that make these systems so interesting despite the implementational problems? One could say there are basically two reasons, or applications, where FH is of great importance— resistance to adverse propagation conditions, and its capability to withstand larger amounts of interference (accidental or intentional).

Using a large number of frequencies makes a well-designed FH system highly resistant to narrowband frequency selective fading. In such a fading environment, certain frequency channels will be exposed to deep fades, whereas most of the other channels will work well. A frequency hopper will be subject to these deep fades now and then, but will never stay long in such a fade. In fact, in a Raleigh fading environment with reasonably average power, a vast majority of the frequencies will provide adequate signal power. Combining frequency hopping with error correction coding will be highly capable of correcting the errors occurring when the system hits the fading minimum. This technique is extremely useful, in particular, when considering mobile communication systems with slowly moving stations. In a narrowband system, when the receiver becomes stationary and happens to find itself in a fading minimum, error correction techniques described in the previous chapter are practically useless. The decoder will hardly receive any symbols of adequate quality to correct the erroneous ones. Here, frequency hopping will make a stationary receiver "move around" in the standing wave pattern around him. In addition, if, in a fast FH system, the chip duration is small compared to the delay spread (e.g., in FH systems for the HF range), the receiver will hop to another frequency before the delayed multipath components have the chance to reach the receiver. In this case, the receiver effectively "hops away" from the intersymbol interference.

Example 2.3 Frequency Hopping in Rayleigh Fading

A binary digital radio link utilizes DPSK modulation and frequency hopping. The transmission is disturbed by Gaussian noise and successive bits can be assumed to be received with independent Rayleigh fading amplitudes. A bit error probability of 10^{-4} is required.

a. What is the required SNR if no coding is used?
b. What is the required SNR if a simple single error correcting Hamming (15,11) code is used?

Solution:

a. The received power (SNR) in a Rayleigh fading channel [7] is exponentially distributed, that is,

$$p(\gamma) = \frac{1}{\gamma_0} e^{-\gamma/\gamma_0}$$

where γ_0 is the average SNR. Now, the bit error probability for a DPSK link given a constant SNR γ, is

$$P_e(\gamma) = \frac{1}{2} e^{-\gamma}$$

Combining these two allows removing the conditioning on γ:

$$P_e = \int P_e(\gamma) p(\gamma) d\gamma = \int \frac{1}{2} e^{-\gamma} \frac{1}{\gamma_0} e^{-\gamma/\gamma_0} d\gamma = \frac{1}{2(1 + \gamma_0)} \approx \frac{1}{2\gamma_0}$$

Requiring a P_e of 10^{-4} yields an average SNR $\gamma_0 = 5000$ (37 dB)

b. The code can correct one error in 15 transmitted bits. The code word error probability for low error probabilities becomes:

$$P_{cw} \approx \Pr[\leq 1 \text{ error in codeword}]$$
$$= (1 - P_e)^{15} + 15 P_e (1 - P_e)^{14} \approx 14 \cdot 15 P_e^2$$

Using the common approximation [7]

$$P_e' \approx \frac{d_{min}}{n} P_{cw} \approx \frac{3}{15} 14 \cdot 15 P_e^2 = 42 P_e^2$$

Using the result from the a) part and noting the fact that only 11/15 of the energy is spent on the transmission of information bits we get

$$P'_e \approx 42 P_e^2 = 42 \left(\frac{15}{2 \cdot 11 \gamma_0} \right)^2 \approx \frac{20}{\gamma_0^2}$$

Requiring a P_e of 10^{-4} yields an average SNR $\gamma_0 = 450$ (26 dB)—a gain of more than 10 dB.

The powerful error correction that can be used in the FH system also has another application. It is possible, for instance, to allow more users than there are frequencies into the available bandwidth. By doing so, collisions, that is overlapping chips, are inevitable. The signals in such a system will no longer be orthogonal (2.7). However, if the excess number of transmitters is moderate, a particular receiver will be hit only now and then, and error correction coding may still be able to recover the original message transmitted. The advantages of such a technique are obvious: We could allow more users into the systems at the price of a moderate performance degradation. Unlike the orthogonal schemes, this multiplexing technique has no definite upper limit on the number of users. Instead the maximum number of transmitters will be determined by the required reception quality.

This capability to withstand interference has been a feature of great interest in military communication systems. In these applications, a hostile party in several ways threatens a communication link. In particular, the enemy may choose to deliberately transmit signals, so-called *jamming signals*, with the explicit purpose of disrupting the transfer of information in the link. Consider the scenario in Figure 2.13 where a link has been established between the transmitter A and the receiver B. A hostile jammer, J, observes the signals from A, and based on these observations tries to transmit signals in order to make the reception in B of the wanted signals as difficult as

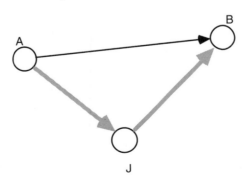

Figure 2.13 Jamming scenario.

possible. Clearly, if A uses an FDM or TDM scheme, or even slow frequency hopping, J would be able to detect the signals from A and immediately transmit a signal on the same frequency. However, if the FH system hops fast enough, the receiver B will already be at some other frequency when the jamming signal hits it. Let τ_{AB}, τ_{AJ} and τ_{JB} denote the path delays between the stations in Figure 2.13, we see that the condition for this to happen is

$$(\tau_{AJ} - \tau_{JB}) - \tau_{AB} \geq T_c$$

where T_c denotes the chip duration. In this case the jammer cannot rely on the observations of the signals from A but has to guess where A will be transmitting the next time. If the hop sequence is such that it appears to be random with every frequency equally probable, the jammer is completely in the dark and may as well randomly jam as many frequencies as possible. This is illustrated in the example below.

Example 2.4 Partial Band Jamming

A frequency hopping system hopping over random L frequencies is being jammed by a so called *partial band jammer*. This jammer randomly selects a fraction q of the frequencies and transmits a jamming signal concentrating all its jamming power on these frequencies. For the sake of simplicity, assume that if the signal to interference ratio (SIR) at the receiver drops below γ_0, the chip is lost (bit error probability $P_b = 1/2$), otherwise the chip is received perfectly ($P_b = 0$). Assume that the wanted signal energy per bit is E_b and that the jammer has energy E_J at its disposal. Estimate the bit error probability as function of the energies and q. Which value q will achieve the maximum bit error probability?

Solution:

Since the jammer distributes its energy evenly over qN frequencies, the SIR at the receiver becomes

$$\Gamma = \frac{qNE_b}{E_J}$$

If Γ is below γ_0 no errors will occur, otherwise a fraction of q of the symbols will be hit and received with $P_b = 1/2$. We can express this as

$$P_b = \begin{cases} \dfrac{1}{2}q & \dfrac{qNE_b}{E_J} = \gamma_0 \\[2ex] 0 & \dfrac{qNE_b}{E_J} < \gamma_0 \end{cases}$$

We can clearly see that the jammer should choose q such that the SIR falls just barely below threshold,

$$q^* = \min\left(1, \frac{\gamma_0 E_J}{NE_b}\right)$$

and the corresponding error probability becomes

$$P_b^* = \min\left(\frac{1}{2}, \frac{\gamma_0 E_J}{2NE_b}\right)$$

The bit error probability decays inversely proportional to the wanted signal energy (Raleigh fading). This result holds also when using a more detailed model to describe the bit error probability as a function of the SIR.

The bandwidth expansion factor N is usually called the processing gain of the system. It can be seen from the final expression in the example that the system achieves the same performance as a single channel system with a transmitter power that is N times larger than in the frequency hopping system.

The choice of hop sequence clearly depends on the application. In the jamming example, the hop sequence has to appear randomly, that is, be impossible to predict for a jammer or eavesdropper. For the civilian application, as a countermeasure against fading, this does not seem to be critical, as long as all frequencies are used regularly. In a system with N frequencies and a hop sequence length of L time slots, there are

$$M = (N!)^L \tag{2.9}$$

different orthogonal sequences. The number of feasible sequences grows very fast with the sequence length. In civilian systems with no hostile interference short sequences that are repeated cyclically are well suited to do the job. Such a short sequence may be easily detected, and future frequencies can be readily predicted which make the synchronization process fast and reliable.

In military applications however, the sequences have to be long and hard to decipher. This requires a large L. If such a sequence is chosen and communicated secretly to the receiver, it would be virtually impossible for the jammer to predict the next chip-frequency. There are two practical problems involved with this. First, synchronization becomes slow and complex. In the worst case, we would have to wait one full hop sequence cycle (or in noise even more!) before transmission could start. The other problem involved with long sequences is that we would need a compact way of describing which one of the M sequences that we have chosen. Just enumerating all M sequences requires a number of $L \log_2 (N!) = L (N/2) \log_2 N$ bits. To give a realistic example, if $L = 1000$ and $N = 100$ there would be 330 000 (!) bits required to fully specify the hop sequence. For obvious reasons, in practice only those subsets of all feasible hop sequences that have compact descriptions are used. These are called *pseudo noise* (PN) sequences that are generated by linear feedback shift registers (LFSRs). We will study some of the properties of these sequences in the next section.

Direct Sequence Systems

Direct sequence (DS) modulation represents the other classical spread spectrum technique. DS systems use long and complex, but usually binary waveforms typically of the shape,

$$u_i(t) = \sum_{k=1}^{N} c_{ik} p(t - k\tau) \qquad c_{ik} \in \{+1, -1\} \qquad (2.10a)$$

where

$$p(t) = \sqrt{E_0/N} \qquad 0 \le t \le \tau \qquad (2.10b)$$

that is, a rectangular pulse of duration τ. The name "direct sequence" is derived from the fact that the data is directly and antipodally (PSK) modulated on the pulse train, the code or code sequence $c_i = \{c_{ik}\}$.

The transmitted signal $s_i(t)$ could thus, according to (2.1), be written as

$$s_i(t) = a_i u_i(t)$$

The time interval $\tau = T/N$ is also here denoted a "chip" (Figure 2.14). N is usually a large number, which yields a short chip duration and thus a

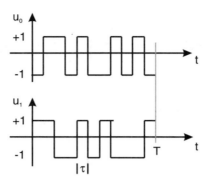

Figure 2.14 Example of signals in DS systems, $N = 11$.

large bandwidth. The signal in (2.10) has a *chip rate* $1/\tau$ that exceeds the information rate $1/T$ by a factor N (i.e., bandwidth expansion factor N).

Deriving the scalar product of two such signals yields

$$\mathbf{u}_i \cdot \mathbf{u}_j = \sum_{\kappa=1}^{N} c_{ik} c_{jk} p^2(\tau - kt) = \frac{E_0}{N} \sum_{\kappa=1}^{N} c_{ik} c_{jk} \qquad (2.11)$$

The last sum is the cross correlation of the two codes c_j and c_i (Example 2.1). To achieve orthogonal signaling, this sum has to be zero for different i and j.

In many DS systems with high chip rates and long code lengths, the synchronization is difficult since variations in propagation delay make it virtually impossible to synchronize all users. In addition, delayed versions of the original signal may be received due to multipath. To avoid the multipath problem and to ease synchronization, one aims at designing sequences that are self-orthogonal, that is, where all cyclically shifted version of the sequence are orthogonal. Synchronization now in principle becomes simple—the optimum receiver would consist of a bank of correlators, each comparing the received signal with a delayed version of the signal. The code delay in the correlator with the winning output would be a measure of the time delay. When synchronization is established, only the output of the winning correlator will be used for the demodulation of subsequent symbols. This fact leads us to a simpler synchronization scheme—the sliding correlator indicated in Figure 2.15. Here a single correlator is used in which the sequence time offset (delay) is slowly varied (slid) between receiver and transmitter. Running the sequence generator at a slightly higher or lower chip rate can do this. Due to the self-orthogonality property, the correlator

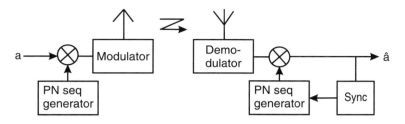

Figure 2.15 DS spread spectrum system.

will not produce any output (but noise) until the correct delay is reached. When a signal of sufficient strength is reached, the correlator is locked.

Being very convenient for synchronization, the self-orthogonality property of a code has a serious drawback. From the definition of the u_is in (2.10) it is clear that all the signals of this type can be described in a vector space of (at most) N dimensions (e.g., use $\phi_i(t) = p(t - i\tau)$ as base functions). Obviously there cannot be more than N orthogonal wave forms in this vector space. By using a self-orthogonal code, that is, a code where all N cyclic shifts are orthogonal, we effectively use up all dimensions. Shifted versions of the other signals $u_i(t - j\tau)$ can thus not be orthogonal to u_i. In practice one will therefore have to trade off synchronization properties against interference (cross-correlation) properties.

Going towards practical code design, let us first study the interference properties. Let r_{ijl} denote the cross-correlation between signal $u_i(t)$ and $u_j(t - l\tau)$, that is, at time shift l. Then

$$r_{ijl} = \frac{1}{N} \sum_{\kappa=1}^{N} c_{i(k+l)} c_{jk} \qquad (2.12)$$

The target is to make the r_{ijl} to be as small as possible. As was noted above, there are no codes with $r_{ijl} = 0$ (except when $i = j$ and $l = 0$), that is, codes that are both orthogonal and self-orthogonal. In fact, we have the following bound

$$\max r_{ijl} \geq \sqrt{\frac{M-1}{MN}} \approx \frac{1}{\sqrt{N}} \qquad \text{(Welsh bound)} \qquad (2.13)$$

where the last approximation holds for large signal sets (i.e., large M). The largest cross correlation coefficient thus decays only as the square root of the code length. There are in fact classes of sequences (so called Kasami and

Gold sequences) for which the bound (2.13) is also an upper bound on the cross correlation. This cross correlation can, however, still be quite high. In a system with N = 100, we would have max $r_{ijl} \geq 0.1$, i.e. 0.1^2 = 1% of the signal power will hit some other user.

A class of sequences with more favorable properties would be the purely *random sequences*. Such (binary) sequences $X_i = \{X_{i1}, X_{i2}, \ldots X_{iN}\}$ would be generated by a sequence of uncorrelated, balanced coin tosses. Every chip in this sequence would take the values +1 and −1 with equal probability independent of other chip values. Calculating the autocorrelations yields

$$R_{ijl} = \frac{1}{N} \sum_{\kappa=1}^{N} X_{(k+l)} X_{jk}$$

Note that R_{ijl} is a random variable with

$$E[R_{ijl}] \approx 0 \qquad (\text{unless } i = j \text{ and } l = 0)$$

$$Var[R_{ijl}] \approx 1/N \qquad (\text{unless } i = j \text{ and } l = 0)$$

For long sequences, due to the law of large numbers, these random sequences will have the wanted properties. Due to the central limit theorem the R_{ijl} will have roughly Gaussian distribution. A reasonable interference model is that the users affect each other as Gaussian noise with a relative power that is N times lower than the wanted signal. However, for moderate N, the variance of the cross correlation is large, and a coin tossing process may produce a sequence with bad properties.

Random sequences are of course not very practical to generate. The amount of information describing the sequence that has to be communicated to the receiver is simply too large (actually the entire sequence would have to be shared by the receiver and transmitter). A class of sequences that retains the favorable correlation properties, but are easy to describe, are the so-called Pseudo Noise (PN) sequences generated by Linear Feedback Shift Registers (LFSR) (Figure 2.16).

The LFSR is a (binary) shift register where the delay element outputs are weighted and summed. The result is fed back to the input of the register. Note that the delay element contents, weight coefficients, and the sum are taken over the field of binary numbers, that is, $g_i \in \{0, 1\}$, and the sum is a modulo-2 summation. An LFSR is an autonomous state machine that can be described by the state diagram in Figure 2.17. Since the network has no input signals, the next state is determined by the previous state. Typically

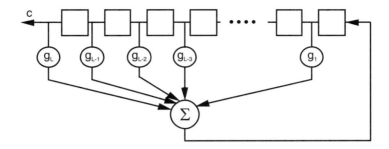

Figure 2.16 Linear feedback shift registers (LFSR).

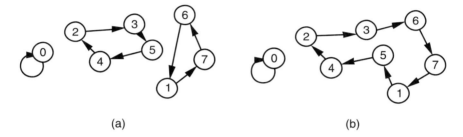

(a)   (b)

Figure 2.17 Sample state diagrams of LFSR: a) Multiple sequences, b) maximal length sequence.

the LFSR has a state diagram that contains multiple loops as in Figure 2.17a. However, there are settings of the feedback coefficients that yield a state diagram as in 2.17b, where there are only two cycles—one containing the zero state and one containing all the other states. Given that the register is initially in a nonzero state, it will run through all states exactly once. That the output (consisting of a sequence of register contents) will run through all L-tuples (except 0000..000) of binary symbols may not come as a surprise. Such a sequence is termed a *maximal length sequence* or just *m-sequence*.

Since there are 2^L states, the m-sequences have length $N = 2^L - 1$. For reasonable register lengths, there are quite a large set of such sequences, all with correlation properties similar to the random sequences. The number of parameters to describe this sequence is just $2L$, the feedback coefficients g_i and the initial register state.

The autocorrelation properties are shown in Figure 2.18. After the correlation peaks corresponding to $l = 0$, the correlation drops to $-1/N$. The sequence is thus for large N almost self-orthogonal. The cross correlation properties resemble those of random sequences.

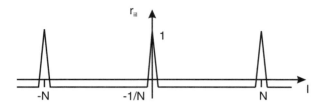

Figure 2.18 Autocorrelation of m-sequence as function of time shift *l*.

The favorable synchronization accuracy of direct-sequence systems have other interesting applications. In fact, the path delay (or to be precise, changes in the path delay) can be measured with high accuracy, that is, within ±1 chip. Several modern navigation systems, including the satellite based Global Positioning System (GPS) utilize this technique.

The synchronization accuracy can also be used to resolve multipath propagation and to estimate the channel impulse response. In a multipath environment, there will not only be one signal component present, but several delayed versions. If the path delay differences are larger than the chip duration, our synchronization scheme could lock onto any of these paths, thereby completely eliminating any multipath fading. The output of the sliding correlator may look something like Figure 2.19. As the time shift is varied, several peaks in the output signal, each proportional to the signal envelope corresponding to that multipath delay, will occur. In fact, Figure 2.19 represents an estimate of the (magnitude of the) instantaneous impulse response of the channel sampled at the chip intervals. To be more precise, the sample points represent the (vector) sum of all the signal components integrated over the chip symbol period. If the number of multipath components is large in every chip bin (poor resolution), this sum will exhibit a rather large variation—a popular model is to assume that the distribution

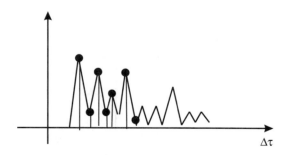

Figure 2.19 Sliding correlator output in synchronizing circuit of DS system.

of the sum amplitude is a Rayleigh distribution. If the number of components is very small (typically only one or no component in each bin) the variations are very limited. There are now several ways of utilizing this property. If we simply lock onto the strongest signal components, we will effectively use the self-orthogonal property of the code to suppress the other components. For m-sequence multi-path components falling into adjacent bins will in this case suffer a $1/N$ power reduction. The effect is a path diversity receiver— in this case using selection diversity [7, 1]. Clearly, instead of locking onto only one of these peaks, one could make use of several multipath components. The effect would be a combining diversity receiver, for example, of the maximum ratio combining type. Since such a receiver would be designed using a tapped delay line, it has been coined a *rake* receiver due to the visual similarity of the block diagram to the garden tool.

As can be seen, DS systems are obviously well suited for digital signal processing and implementation with digital logic. Up until a few years from now, the main use of this technique was in the military sector. Systems with long sequences that are hard to predict are also very hard to jam. The jammer has to rely on rather blunt jamming techniques such as transmitting wideband noise. Another advantage is the noise-like structure and the very low spectral density of the signal (approximately N times lower than the original signal). Such signals are hard to detect, in particular with narrow band receivers (so-called low probability of intercept (LPI) systems).

Signal management is very easy in DS-CDMA systems. Users can simply pick their codes at random. Since the code set is usually very large, the risk of a collision is low. The main practical drawback of the DS-systems emanates from the same feature. Since signals are not orthogonal, a small number of users will interfere with each other. Although the relative cross correlations can be made small, the large dynamic range of the radio channel can cause severe problems. Assume for instance that a sequence of lengths $N = 100$ is used. In such a system users will create interference to each other that is 100 times or 20 dB lower than the wanted signal. However, if the interfering transmitter is very close to the receiver, whereas the wanted signal comes from far away, the interferer may get a power advantage that may be by far more than these 20 dB, maybe up to 100 dB. In this case, reception of the weaker signal is not possible. This problem is treated further in Chapter 9.

2.2 Link Performance Models

As was seen in Section 2.1, the actual calculation of a transmission quality measure such as the bit-error probability in a multi-user scenario may be quite complex. The performance will depend on many parameters such as the waveforms, the instantaneous amplitudes, and the phases of the interfering signals. When analyzing wireless systems with a large number of terminals and complex propagation conditions, the exact analysis will pose a formidable task. Thus, some simplified model or procedure is necessary. In Example 2.1, however, it was seen that the signal-to-interference ratio, E_0/E_1, played a key role in determining the bit error probability. Using this quantity will be the approach in this book. One transmitter-receiver pair, one link, at a time will be studied and all the interfering signals will be characterized by their aggregate, or sum power, that is,

$$\text{Quality} = f(E_0/E_1) = f(\Gamma)$$

where Γ denotes the signal-to-interference ratio. In those cases where the receiver noise cannot be neglected, the Noise energy has to be included in the interference term, and Γ is referred to as the signal-to-interference + noise ratio (SINR). There are cases where this is indeed a reasonable assumption, as we have already seen in Section 2.1. One such case is when the number of interferers is large, and when all these inteferers have similar received powers. In this case the central limit theorem will ensure that the total interference signal approaches a zero mean Gaussian vector, which is characterized only by its energy N_0. If the discussion is confined to the bit-error probability as performance measure, this quantity can straightforwardly be computed by means of classical results from Gaussian detection theory.

This is illustrated by Figure 2.20, which shows the bit-error probability for a coherent BPSK reception as function of the total SIR if the interference is composed of N BSPK signals of identical magnitudes, but random (uniform) phases. As can be seen, already when there are three interferers, the Gaussian approximation is very good.

More complex is the situation when the transmission channel exhibits frequency selective multipath propagation. Clearly, the degree of multipath (delay spread, Doppler frequency, and so forth) will influence the performance and, for example, the SINR will not be the sole quantity that determines the bit-error probability. On the other hand, most communication systems designed for mobile communications will have some means of combating multipath fading. This could be a combination of error control coding, frequency hopping (FH), or equalization techniques, or a Rake-receiver in the DS-spread spectrum case.

In Figure 2.21, one example from the GSM system is shown (which uses a combined FH/equalizer/coding scheme). It is notable that, although

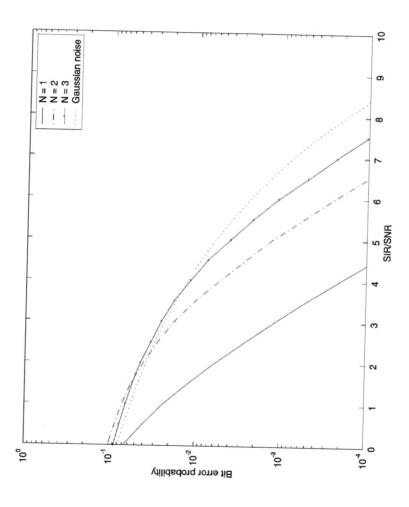

Figure 2.20 Bit-error probability for BPSK as function of signal-to-interference ratio when interference is composed of *N* = 1, 2, 3 identical BPSK signals with random phase (see Example 2.1). Dashed line BPSK performance in Gaussian noise with corresponding SNR.

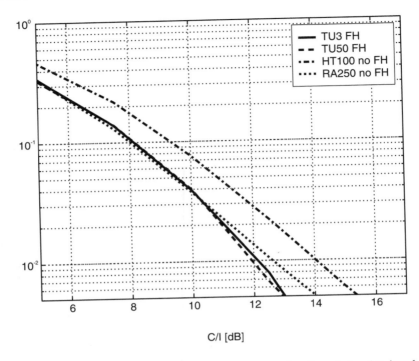

Figure 2.21 Block-Error probability GSM/EDGE system as function of signal-to-interference for various multipath propagation conditions (courtesy Ericsson Radio Systems AB).

the multipath channel characteristics are varying considerably between these cases, the general trend is the same. The same type of behavior can be found also after equalization and error control decoding. Figure 2.22 gives an example of such a model for the packet data communication mode in GSM, GPRS. Here, four different coding schemes may be used, each of them providing a certain transmission quality (here the data block error probability) as a function of the instantaneous SIR.

Graphs like 2.21 also illustrate very well the strategy for analyzing complex wireless networks that will be utilized in the following chapters. In such a network, a large number of transmitters may be active and their signals propagate to the receivers (intended and unintentional) over diverse propagation paths. The detailed analysis of such networks is a formidable task. Instead, the network analysis will be simplified to computing only the wanted signal-to-interference ratios at the receiver (Figure 2.23). We will then map these SIR values on the quality measures of interests using graphs

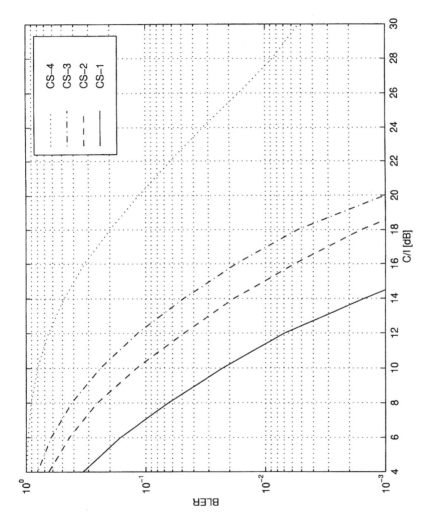

Figure 2.22 Block-error rate as function of SIR in GPRS for 4 different coding schemes [9].

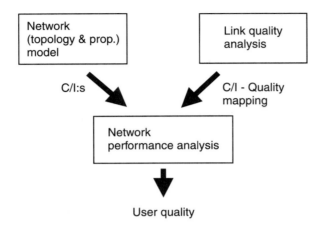

Figure 2.23 Simplified network analysis strategy.

as shown in Figure 2.22, which are derived from the analysis of a single link. This link analysis per se may be quite complex and may include extensive simulation studies. The latter is not a problem since once we have done the calculations and the mapping is known, evaluating the performance given the SIRs becomes a simple table reference.

Problems

2.1 A two-user system utilizes BPSK signaling. The two users are using the waveforms

$$s_0(t) = a_0 \sqrt{\frac{2E_0}{T}} \cos\left(2\pi \frac{t}{T}\right) \qquad 0 \le t < T$$

and

$$s_1(t) = a_1 \sqrt{\frac{2E_1}{T}} \cos\left(2\pi \frac{t}{T}\right) \qquad 0 \le t < T$$

The data symbols $a_i \in \{-1, +1\}$ are independent and equally probable. The signal $s_0(t)$ reaches our receiver on a direct path, whereas the signal $s_1(t)$ is subject to multipath propagation. The received signal may be written as

$$r(t) = s_0(t) + s_1(t) - s_1(t - \tau)$$

Study the system for the signal-to-interference ratios $E_0/E_1 = 2$ and $E_0/E_1 = 1/2$.

a) Assume that a conventional correlation detector is used to detect a_0. Determine the error probability as function of the delay spread τ.
b) The communication link is also disturbed by additive white Gaussian noise with spectral density $N_0/2$. Determine again the error probability as a function of the delay spread τ if a conventional correlation detector is used to detect a_0.

2.2 A cellular telephone in the NMT900 system allows that the signals can be received at the same time the transmitter is in operation. This is made possible by the use of a Duplex-filter, provided the signals have to be separated in frequency by 10 MHz. Assume that the phone is in operation at the cell border, 10 km from the base station. The phone may operate properly down to SIR of 20 dB. Assume that free space propagation prevails and that the base station uses the same power as the mobile. Both stations use antennas with an antenna gain of 3 dB. Determine the minimum attenuation of the duplex filter in order to make reception possible at the mobile. What would happen if the propagation loss would be 20 dB higher than free space?

2.3 A binary shift register, according to the figure below is used to generate a synchronization sequence. The binary {0,1} output of the register is coded antipodally according to the mapping in the figure below.
a) Show that this register generates a maximum-length sequence, provided it is not initially in the all-zero state.
b) Determine the autocorrelation of the output sequence!

2.4 Show also that the LFSR in the figure below generates an m-sequence and determine the cross-correlation with the sequence in Problem 7.3.

2.5 A TDMA system uses an m-sequence of length 15 bits for synchronization purposes. In order to find the sync sequence and to receive the frame correctly, all 15 sync bits have to be received correctly. The radio channel can be modeled as a BSC with bit error probability $p = 5\%$.
a) Compute the "miss-probability" of the synchronizer, P_m, i.e., the probability that the receiver will not find the sync sequence due to bit errors even though the system is in sync.

b) Estimate the probability of false sync, P_f (i.e. the probability that some random data or a randomly shifted version of the sequence appears as a correct sequence). Assume that data bits appear correctly with probability 1/2.

c) Repeat a) and b) and compute P_m and P_f if the receiver finds sync if there are at most 3 errors in the sync sequence.

2.6 An FH-system with N = 100 consecutive frequency channels using DPSK-modulation is subject to jamming. The jammer decides to use a partial band jamming technique and selects K frequency channels in which it transmits white Gaussian noise. The jammer can generate 10 times more signal energy at the receiver than the legal transmitter. The same jamming energy is used in all K selected channels. Determine the K that maximizes the bit error probability of the FH system and calculate this bit error probability.

2.7 A DS cellular system has rather long chip duration compared to the multipath profile duration such that the received powers in each chip bin can be assumed to be independent and exponentially distributed. Assuming that the receiver considers three chip sync positions and can choose between either i) selecting the strongest component, or ii) the use of three branch rake receivers using maximum ratio combining. Compare the resulting average SNR for the two receiver strategies i) and ii) if the average branch-SNRs are

a) 10, 10, 10 dB;

b) 15, 10, 5 dB.

References

[1] Proakis, J. G., *Digital Communications*, New York: McGraw-Hill, 1995.

[2] Hagerman, B., "Downlink Relative Co-Channel Interference Powers in Cellular Radio Systems," *Proc. IEEE Veh Tech Conf., VTC '95*, Chicago, July 1995.

[3] Hagerman, B., "Strongest Interferer Adaptive Single-User Receivers in Cellular Radio Environments," *Proc. Globecom '95*, Singapore, Nov. 1995.

[4] Lupas, R., and S. Verdu, "Near-Far Resistance of Multi-User Detectors for Asynchronous Gaussian Multiple-Access Channels," *IEEE Trans. Comm.*, Vol. COM 38, April 1990.

[5] Woozencraft, J. M., and I. M. Jacobs, *Principles of Communication Engineering*, New York: Wiley, 1965.

[6]   Mouly, M., and M. B. Pautet, *The GSM System for Mobile Communications*, published by the authors, 1992.

[7]   Ahlin, L., and J. Zander, "Principles of Wireless Communication," Student Literature, 1997.

[8]   Wigard, J., et al., "BER and FER Prediction of Control and Traffic Channels for a GSM Type of Air-Interface," *48th IEEE Vehicular Technology Conference*, 1998, Vol. 2, pp. 1588–1592.

[9]   Schramm, P., et al., "Radio Interface Performance of EDGE, a Proposal for Enhanced Data Rates in Existing Digital Cellular Systems," *48th IEEE Vehicular Technology Conference*, 1998, Vol. 2, pp. 1064–1068.

3

Wireless Network Models

3.1 The Resource Management Problem

In a wireless mobile communication system, mobiles move around in the service area and will, from time to time, require communication services in the form of a wireless connection to the fixed network as outlined in Chapter 1 and Figure 1.4. As was noted in Chapter 1, like most resource allocation problems, this one has two distinct aspects, a provider perspective and a consumer perspective. The network provider (operator) owns and provides the communication resources to the users who consume them in exchange for money. The two parties have different interests: The operator wishes to run a profitable business, whereas the user is willing to pay for the quality services he appreciates. In the simplest case where all users are provided with the same service offering (for the same pay), the revenues of the operator will be maximized if he can maximize the number of users using his system. The latter certainly depends on economical factors such as the price and competing operators or services, but also on the technical limitation of the systems. The maximum number of users that can be served by the system is here called the capacity of the system. The user in turn is interested in getting the best Quality-of-Service (QoS). It is obvious that, services requiring more resources per user, will limit the capacity of the systems and that in general the capacity will be a function of these QoS requirements. Since the operator is paying for the system, it is natural that the design process is ruled by his perspective. The problem formulation will thus be to design systems where the number of users is maximized for a given QoS requirement. The latter will be the network provides service offering and will be the constraint in our optimization problem. This section will provide a more formal definition of this problem. We will see that it has two aspects:

1. The wireless network design problem:
   Here the problem of designing the fixed network infrastructure is considered. How many RAP:s are required? Where should they be placed? What fixed network capacity has to be provided for the different RAP:s? This planning problem will be addressed more in Chapter 11.

2. The radio resource allocation problem:
   Given a certain infrastructure design (RAP locations), how should the wireless resources be allocated to meet the instantaneous demand of the users and mobile terminals moving around in the network? This is the problem addressed in this section.

A mobile terminal requiring a connection will be called an *active* terminal. The system will, whenever possible, attempt to establish a unique two-way link between an active terminal and some access port in the network. Assuming that, at a certain instant, there are M active terminals in the service area numbered (in some simple way) using the set

$$\mathbf{M} = \{1, 2, 3, \ldots M\}$$

As the number of active terminals will change over time, the size of this set M, will vary. Assume further that M is a stochastic variable. The distribution of M will depend on the user behavior, that is, traffic (call) intensity and the duration of a user session. In addition, the distribution will also depend on the system behavior, for example, to which degree the service requirements of the individual users can be met.

In this section a simple model for user behavior and QoS requirements will be introduced. Users will be assumed to initiate and terminate sessions according to some random processes. During these sessions the users may request various services characterized by various sets of QoS parameters (e.g., bit rate, error probabilities, delays, and so forth). The QoS concept is discussed in some more detail in Section 3.2. The problem is approached in several ways. First, we will try to establish how these services can be provided to the users at some instant of time. This is referred to as a *snapshot analysis*. This type of approach is of direct relevance to all kinds of circuit switched service where the users require physical access to the radio medium at all times (and we will thus need to fulfill the QoS requirements at virtually any snapshot). Nonreal time services like packet-switched data services will be treated mainly in Chapter 10.

In the following snapshot analysis, it is assumed that the active terminals are uniformly distributed over the service area, that is, the probability density of the location vector $R = (X, Y)$ of some given active terminal is constant

$$p_{XY}(x, y)dxdy = \Pr[X \in [x, x + dx], Y \in [y, y + dy]] = \frac{1}{A}dxdy$$

where A is the area of the service area (see definition in Chapter 1). A given terminal is thus equally likely found in any part of the service area. If the number of active terminals is large and the terminals become active independently of another, point process theory indicates that the locations of active terminals form a two-dimensional Poisson point process. A two-dimensional Poisson process is exactly characterized by the fact that the number of points (terminals) in disjoint areas are independent Poisson distributed random variables. Clearly, given that a number of active terminals are observed in a given area, the locations of these terminals will be independent and uniformly distributed over that particular area. Let us denote the intensity (rate) of this process by ω (active terminals/area unit). As will be seen, the terminals do not act completely independently since they share (compete for) the same set of signals. Still this may serve as a reasonable approximation if the traffic load is not too high, that is, when most terminals will be assigned signals (channels). The relationship between the frequency of sessions and the area A on one side and ω on the other is not trivial and will depend both on the behavior of the terminals as well as the behavior of the system (e.g., how sessions are handled).

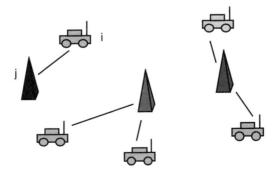

Figure 3.1 The Radio Resource Management problem in a wireless network.

The terminals are served by access ports (base stations), numbered from the set

$$\mathbf{B} = \{1, 2, 3, \ldots B\}$$

Now, assume that there are C waveforms numbered from the set

$$\mathbf{C} = \{1, 2, 3, \ldots C\}$$

are available for establishing links between access ports and terminals. Further, let θ_{ij} denote the normalized cross correlation between signal i and j as defined in Example 2.1. To establish radio links in the system each terminal has to assign:

- An access port from the set **B**;
- A waveform from the set **C**;
- A transmitter power for the access port and terminal station.

This assignment (of access port, channel, and power) is performed according to the resource allocation algorithm (RAA) of the wireless communication system. The assignment is restricted by the interference caused by the access ports and terminals as soon as they are assigned a waveform and when the start using it. Another common restriction is that access ports are in many cases restricted to use only a certain subset of the available waveforms. Good allocation schemes will aim at assigning links with adequate SIR to as many (possibly all) terminals as possible. Note that the RAA may well opt for not assigning a waveform to an active terminal if this assignment would cause excessive interference to other terminals.

Even if the power assignment problem is disregarded (i.e., assume that all transmitters use some given, fixed power), the resource allocation problem is indeed a very complex one. An assignment of this kind is characterized by the fact that terminals and access ports can be divided into the sets

$$\mathbf{M}^{(k)} = \{ j : \text{terminal } j \text{ has been assigned waveform } k \}$$
$$k = 0, 1, 2, 3, \ldots C$$

$$\mathbf{B}^{(k)} = \{ i : \text{RAP } j \text{ has been assigned waveform } k \} \quad k = 0, 1, 2, 3, \ldots C$$

where "waveform 0" has been introduced as a convenient notation for the fact that a terminal has not been assigned any channel at all, that is,

$$\boldsymbol{M}^{(0)}, \boldsymbol{B}^{(0)}$$

are the sets of terminals and access ports without waveform assignment. Note that

$$C; U; k = 0 \ \boldsymbol{M}^{(k)} = \boldsymbol{M} \qquad C; U; k = 0 \ \boldsymbol{B}^{(k)} = \boldsymbol{B}$$

The sets $\boldsymbol{M}^{(k)}$ are usually (but not necessarily) disjoint (each terminal uses only one waveform at the time) whereas the set $\boldsymbol{B}^{(k)}$ usually is not (a RAP may serve multiple terminals using different waveforms). Even for very reasonable numbers of terminals and base stations, the number of possible waveform and access port assignments is astronomical. The following exercise illustrates this fact.

The interference constraints on resource allocations will now be studied in somewhat more detail. According to the analysis approach outlined in the previous chapter, only the signal and interference power levels in all access ports and terminals has to be computed. This can be done using the *link gains* between transmitters and receivers, i.e. where G_{ij} is the power gain of the waveform on the path between access port i and terminal j. This means that the received power in receiver j, denoted $P_{rx,j}$, can be written as

$$P_{rx,j} = P_{tx,i} G_{ij}$$

where $P_{tx,i}$, is the transmitter power at transmitter j. Further assume that the G_{ij}:s are random variables. Collecting all link gains in matrix form results in a $B \times M$ rectangular matrix, the link gain matrix

$$\boldsymbol{G} = \begin{pmatrix} G_{11} & G_{12} & \cdots & G_{1M} \\ G_{21} & G_{22} & \vdots & \vdots \\ \vdots & \vdots & \vdots & \vdots \\ G_{B1} & G_{B2} & \cdots & G_{BM} \end{pmatrix}$$

The link gain matrix describes the (instantaneous) propagation conditions in the system. Note that in a terminal system both the individual components (terminal motion) and the dimension of the matrix (traffic pattern) may vary over time.

The task of the resource allocation scheme is to find assignments for which the QoS is adequate in as many links as possible (preferably all). In our snapshot analysis this means that the instantaneous transmission quality has to be sufficient in these links. Using our simplified link quality model from the previous chapter, this in turn means that the SIRs (or actually the signal-to-interference+noise ratio) are larger than some requirement which could possibly be different for different links. This means that, given that a terminal j has been assigned to access port i_0 using waveform c_0, the following inequalities must hold

$$\Gamma^u_{i_0 j} = \frac{P_j G_{i_0 j}}{\displaystyle\sum_{m \neq j} P_m \theta_{0,m} G_{i_0 m} + N_{i_0}} \geq \gamma^u_j$$

$$\Gamma^d_{i_0 j} = \frac{P_{i_0} G_{i_0 j}}{\displaystyle\sum_{b \neq i_0} P_b \theta_{0,b} G_{i_0 b} + N_j} \geq \gamma^d_j \tag{3.2}$$

where $\Gamma^u_{i_0 j}$, $\Gamma^d_{i_0 j}$ denote the SIR in the up (terminal-to-access port) and down (access port-to-terminal) link of the connection. N_j and N_{i_0} denote the receiver (thermal) noise power at the terminal and access port respectively. The θ_{ij} denote the cross-correlations between waveforms i and j. The effective interference power received at terminal m from access port b is thus $P_b \theta_{m,b}$. The total interference power is assumed to be the sum of these effective powers of the individual interference components.

Orthogonal signal sets:

An important special case of engineering relevance is when C consists of orthogonal waveforms only, that is,

$$\theta_{ij} = \begin{cases} 1 & i = j \\ 0 & i \neq j \end{cases}$$

Waveform sets of this type are the ones used in classical, first generation mobile telephony systems. Here the waveforms are FDMA waveforms, i.e. nonoverlapping narrowband signals or channels. The assignment problem is in the case referred to as the channel allocation problem. The constraints (3.2) simplify to become

$$\Gamma^u_{i_0 j} = \frac{P_j G_{i_0 j}}{\displaystyle\sum_{\substack{m \neq j \\ m \in M^{(c)}}} P_m \theta_{0,m} G_{i_0 m} + N_{i_0}} \geq \gamma^u_j$$

$$\Gamma^d_{i_0 j} = \frac{P_{i_0} G_{i_0 j}}{\displaystyle\sum_{\substack{b \neq i_0 \\ b \in B^{(c)}}} P_b \theta_{0,b} G_{i_0 b} + N_j} \geq \gamma^d_j$$

where the interference summations now have to be taken over all terminals (RAP:s) using the same channel c only.

Note that there is no guarantee that it is possible to comply with all the constraints (3.2) for all the M terminals, in particular if M happens to be a large number. A system designer may have to settle for finding resource allocation schemes that assign waveforms with adequate quality to as many terminals as possible. The largest number of users that may be handled by the systems is a measure of the system capacity. Since the number of terminals is a random quantity, and the constraints (3.2) depend on the link gain matrix, that is, on the relative position of the terminals, such a capacity measure is not a well-defined quantity. In order to arrive at a more precise definition of an adequate performance measure, let us recollect that the number of active calls is the random number M and that the link gains in the matrix **G** can be modeled as stochastic variables. Assume that at some given instant, our RAA has succeeded to provide waveforms that provide adequate quality (i.e., satisfying the constraints 3.2) to Y of these terminals. Y will of course also be a stochastic variable. Let us by Z denote the remaining number of terminals, the assignment failures, that is,

$$Z = M - Y \tag{3.3}$$

The assignment failure rate ν was defined as

$$\nu = \frac{E[2]}{E[M]} = \frac{E[2]}{\omega A} \tag{3.4}$$

This quantity is a measure of the average proportion of the allocation scheme that has been successful in providing the terminals with links of adequate quality. For moderate to large ωA, ν is also a good approximation of the probability that it is not possible to provide a combination of waveform and access port to some randomly chosen active terminal at some given instant without violating the constraints (3.2). The instantaneous capacity $\omega_*(\nu_0)$ of a wireless system is the maximum allowed traffic load in order to keep the assignment failure rate below some threshold level ν_0, that is,

$$\omega_*(\nu_0) = \{\max \omega:\ \nu \leq \nu_0\} \tag{3.5}$$

The capacity is thus expressed in terms of active terminals/area unit.

There are many ways to design resource allocation algorithms (RAAs). As illustrated by Figure 3.2, an RAA uses information (through measurements) about the traffic (active users and their QoS requirements) and propagation conditions (link gain matrix **G**), to make its decision about the access port association, the waveforms, and the transmitter power assignment. As could be seen above, finding the optimum resource allocation, that is, the waveform, the power, and the access port assignments that maximize Y for a given link gain matrix, is a formidable problem. In fact it has been shown that this problem in this general form belongs to the set of NP-complete problems. For this class of problems no efficient[1] general algorithm that is capable of doing such an optimal assignment for arbitrary link gain matrices and terminal sets is known. Instead, a number of simple heuristic schemes have been proposed (and are used in the wireless systems of today). These schemes are usually characterized by low complexity and by using simple heuristic design rules. The capacity ω_* achieved by these schemes is, as expected, often considerably lower than what could have been achieved by optimum waveform/access port/power assignment.

Note that the considerations so far are dealing with a specific snapshot. As time goes on, propagation conditions may change (due to mobile terminal movement) as well as traffic conditions (terminal activity status and/or QoS

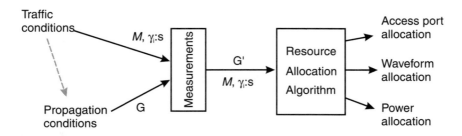

Figure 3.2 Resource allocation algorithms.

---

1. By efficient we mean an algorithm that does not require a number of operations that grow exponentially with the problem size.

requirements). Constantly recalculating and updating the resource assignments to adapt to these changes naturally further increases the complexity and the computational load in an RAA. In addition the measurements taken to reflect the propagation conditions may be both unreliable as well as costly to retrieve. The latter is due to the fact that the measurements normally have to be transmitted on-the-air (on some signaling channel), thereby stealing some of the capacity available for user data transmission.

In practical RAAs, complexity is usually limited by reducing the input data set, either by using partial or aggregate information or by limiting the rate at which the measurements are taken. An example of the former is using measured SIR-values to predict future SIR as a result of RAA actions. The rate at which measurements are taken may vary considerably anywhere between the two extremes:

1. Static resource allocation:
   Here virtually no measurements are taken and the resource allocation is based on a priori statistical information such as the average propagation condition in certain areas (say around a certain access port), the average traffic load conditions and the some given fixed QoS requirement. The resource allocation is mainly done during the planning stage of system deployment.

2. (Ideal) Dynamic resource allocation:
   In this class of schemes, the sampling rate of measurement is high enough to track all changes in propagation conditions and user traffic requirements. Update and reassignments rates are dependent of the rate change in these latter parameters, but 100 times per second or more are not uncommon requirements for such schemes.

Neither of these extremes exists in practice. Certainly no reasonable (almost) static allocation scheme can avoid taking at least some traffic conditions into account (such as which mobiles are currently active). In the same way, the extreme update rates in fully dynamic allocation schemes lead to calculations that are too complex and an excessive signaling load. Dynamic resource allocation schemes are investigated in more detail in Chapter 7.

3.2 Quality-of-Service Models and User Behavior

Although the snapshot analysis provides a basic understanding and establishes reasonable performance limits, the true performance as experienced by the users of the system will be determined by the detailed character of the information exchange and QoS requirements. It has to be understood that the user is interested in receiving a telecommunication service or information service in the wider sense, where the technical systems that are studied here (i.e., the systems that transport bits and messages) are only a part of the service chain (Figure 3.3). Mapping perceived performance on technical parameters in the wireless system is certainly a very complex task. The user experience is influenced not only by the shortcomings of the wireless system but also by other factors such as the performance of the wireline backbone and switching, the service providers application software and hardware, and the user interface provided by the service provider and the terminal manufacturer. In order to allow rational design of telecommunication systems, these overall (end-to-end) requirements are broken down to specific (sub) service requirements for the individual building blocks. Here the interest is focused on studying the behavior of the radio network part of the "transport system." The "services" provided at this level have been coined *bearer services* in the UMTS/3G standardization process. These services have been divided into four classes as outlined in Table 3.1 [1] and are mainly distinguished by their delay requirements ranging from very strict delay requirements in conversational class 2 (e.g., voice services) to the very relaxed requirements in the background-best effort class. In more technical terms, the services in the different classes are characterized by means sets of service parameters, forming a QoS profile. Some of the QoS parameters (service attributes) in the 3G systems are

- *Maximum data rate:* The highest data rate (averaged over some a certain time interval) that a user could expect;
- *Guaranteed data rate:* The lowest data rate (averaged over a certain time interval) that a user is guaranteed;
- *Maximum packet/message size;*
- *Residual bit error rate:* The undetected error rate after the delivery of the information over the service interface;

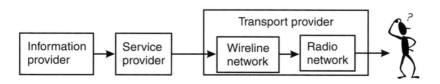

Figure 3.3 Service provisioning in modern information-communication system.

Table 3.1
3G (UMTS) Bearer Service Classes [1]

Service Class	Typical Applications	Service Functional Characteristics
Conversational Real Time (RT)	Voice	• Preserve time relations between entities • Stringent preservation of conversational patterns (low delay)
Streaming RT	Video/Audio streams	• Preserve time relations between entities
Interactive Best Effort (BE)	Web-browsing	• Request-response pattern • Preserve payload (low error rate)
Background BE	File transfer, E-mail	• Not time critical • Preserve payload (low error rate)

- *Transfer delay:* The time delay the packet/message spends between the service access points; This could be a guaranteed delay for every message (e.g., in conversational class services) or defined in statistical terms, for example, average delay or X percentile delay (X% of all messages experience a delay shorter than the specified value);
- *Priority:* Indicates the relative importance of different messages.

In principle an infinite set of QoS profiles and thus different bearer services could be defined by varying these parameters—possibly one combination for each user. In practice, limitations on the number of modulation waveforms, codes etc. restrict the number of service offerings. Most systems will therefore offer a finite set of bearer services, where each parameter will be allowed to take one (out of a few) discrete values. Table 3.2 provides an indication of what ranges these service parameters can take in a 3G wireless system.

In addition to these service parameters, the availability of the services has to be considered since it may vary over time and over the user services. Clearly, users with QoS profiles with large "resource consumption" (e.g., high bit rate, poor location) or with low relative priority will more often experience that the system is not capable of accommodating their service request. The following two quantities to measure the availability of a certain service are introduced:

1. *Service denial probability.* The probability that a user is denied to begin a session with a certain bearer service due to resource shortage.

Table 3.2

Some 3G (UMTS) Service Attribute/Parameter Ranges [1]

Traffic Class	Conversational	Streaming	Interactive	Background
Max bitrate (kbps)	<2000	<2000	<2000—overhead	<2000—overhead
Max SDU size (byte)	<1500	<1500	<1500	<1500
Guaranteed bit rate	<2000	<2000		
Transfer delay (ms)	80—max value	500—max value		
Priority	1,2,3	1,2,3	1,2,3	1,2,3
Residual BER	$5*10^{-2}$, 10^{-2}, 10^{-3}, 10^{-6}	$5*10^{-2}$, 10^{-2}, 10^{-3}, 10^{-4}, 10^{-5}, 10^{-6}	$4*10^{-3}$, 10^{-5}, $6*10^{-8}$	$4*10^{-3}$, 10^{-5}, $6*10^{-8}$

2. *Service interruption probability.* The probability that a user is forced to terminate a session with a certain bearer service due to resource shortage.

Sample Case 1: Voice (Telephony) QoS and Traffic Model

In order to keep the voice quality reasonable, usually residual message (speech phoneme block) error rates have to be lower than 10^{-2} and the end-to-end delay should be fixed and not more than 80–100 ms (preferably lower since other parts of the network such as satellite links may also introduce additional delay). All sessions, "calls," are assumed to have an exponentially distributed duration with average length $1/\mu$ (time units/call). The relative traffic load, $\rho = \lambda/\mu$, describes how many new calls can be expected to arrive during the duration of an ongoing call. The quantity ρ is formally without dimension, but is usually measured in Erlang. The service denial probability is here referred to as the blocking probability (the probability that a newly arrived call is denied service) [2]. The probability that a call will be lost while in progress (the service interruption probability) is referred to as the (call) dropping probability.

Sample Case 2: Web Browsing QoS and Traffic Model

A web browsing session consists of an irregular sequence of file transfers (using the TCP/IP protocol stack). Typical very short messages are transmitted in the uplink from the terminals (corresponding to a mouse-click) a random instant to request rather large files (web pages, pictures, and so forth) to be downloaded into the terminal. This can be seen as a service of the interactive class. The critical QoS characteristic to the user is the response time (i.e., delay between request and the complete reception of the requested page). For large requested files, the delay is dominated by the transfer delay of the files, that is, the average data rate is in fact the critical QoS parameter that will determine the user delay. The undetected error rate at the user level has to be below 10^{-8} corresponding to 1 error in about 10 MB. The radio bearer may however have a larger undetected error probability since the TCP/IP protocol provides end-to-end error control of its own. Classical models for data traffic of this type are based on Poisson distributions—both for the interarrival time of packets as well as the size of the requested files. Recent studies of Internet traffic have however shown that these models tend to underestimate the time between packets and the packet sizes. As alternative, the Pareto distributions, that is, stochastic variables with density function

$$f(x) = \frac{\beta a^{\beta}}{x^{\beta+1}} \qquad a, \beta \geq 0, x \geq a$$

have been proposed [3]. These distributions are heavy-tailed, that is, they assign higher probability to large values of x compared to the exponential distributions found in Poisson point processes. Another approach for simulation purposes, is to use a more detailed model of the actual TCP/IP protocol [4].

After having discussed the user performance perspective, let us now briefly return to the discussion of the network performance. In the previous section, a simple capacity definition was provided (3.5), based on the snapshot concept. When applying this capacity definition, two fundamental problems have to be considered:

1. The QoS—C/I mapping: Is it possible to map the various QoS requirements on to a simple link quality measure? This problem was treated in section 2.4 where it was noted that this mapping could be performed quite successfully. However the temporal aspects, like message delay in non-real time (best effort) services may not be as easily mapped and will require special attention (Chapter 10). Other difficulties involve the temporal aspects related to the variations in service availability due to traffic fluctuations and user mobility.

2. The user behavior and service mix: In order for the definition (3.5) to be precise in the general (multi-service) case requires a model, not only for the number of users, but also for their behavior. What will be the QoS-profile requested for certain user (which is then mapped to the individual) γ_i in the constraints (3.2) and what is the required service availability? Typically, random models will be used for this purpose. A user will, with some given probability, belong to a certain class of users with identical QoS profiles. The probability distribution of class membership is usually referred to as the service mix.

Determining which service mix should be used for the capacity definition is indeed difficult. One approach has been to look at the pricing strategy and maximizing the revenue of the operator. This leads to both optimal service mix and maximal revenue derived from the network. A difficulty is that this type of model disregards the fact that if the demand for the service provided is finite, operators are prone to competition from other similar operators or from alternative technical solutions. In these situations the pricing strategy clearly affects the demand for the services.

Problems

3.1 In the model in Section 3.1, assume that each terminal is assigned only one access port and that each waveform is used only once in each access port. What is the number of feasible assignments?

3.2 In a personal communication system, terminals are distributed according to a 2D-poisson process with intensity parameter ω'. Assume that terminals are active (independently of each other) with probability q.
   a) What is the probability of finding exactly k terminals in the area?
   b) What is the expected number of active terminals in an area of size A?
   c) Given that 5 terminals are observed in an area of size A, what is the probability that more than 3 of these are active?

3.3 In a mobile telephone system, terminals in a certain area are assigned one access port. The access port has in turn been assigned N channels. Calls with exponentially distributed holding times and Poisson arrivals are blocked if all N channels are busy. Show that if N is large, the number of calls in progress in the area in a certain snapshot is approximately Poisson distributed.

3.4 In a wireless network with two access ports and three terminals, the gain matrix is found to be

$$\mathbf{G} = \begin{pmatrix} 0.02 & 0.0005 & 0.05 \\ 0.002 & 0.01 & 0.001 \end{pmatrix}$$

The transmitter power is constant ($=1$) and the noise power N is 0.001 for all terminals. Two (orthogonal) waveforms are available. Determine the optimal access port and channel assignments and the resulting SIRs.

3.5 Repeat problem 3.4, but assume that infinitely many waveforms are available, all having cross correlations

$$\theta_{ij} = \begin{cases} 1 & i = j \\ 0.1 & i \neq j \end{cases}$$

What are the resulting SIRs now?

References

[1] 3GPP Specification TS23.107, Dec. 1999.

[2] Kleinrock, L., *Queueing Systems, Part I: Theory*, New York: John Wiley & Sons, 1976.

[3] Paxson, V., and S. Floyd, "Wide Area Traffic: The Failure of Poisson Modeling," *IEEE/ACM Trans Networking*, Vol. 3, No. 3, June 1995.

[4] Anderlind, E., and J. Zander, "A Traffic Model for Non Real-Time Data Users in a Wireless Radio Network," *IEEE Comm. Letters*, Vol. 1, No. 2, March 1997.

[5] Gudmundson, M., "A Correlation Model for Shadow Fading in Mobile Radio," *Electronics Letters*, Vol. 27, No. 23, Nov. 1991, pp. 2146–2147.

4

Principles of Cellular Systems

4.1 Frequency Reuse

In this chapter we take a closer look at the interference interaction between the different terminals and access points described in the previous chapter. Ultimately, we introduce the first and most basic resource management scheme, static allocation, which is used in all mobile telephone systems of today. In its simplicity, the static allocation scheme is an excellent example demonstrating most of the features and problems in resource management. It is also the point of departure and serves as reference for more advanced schemes.

The starting point will be a system with orthogonal waveforms. To make things interesting, assume now that the number of simultaneous connections (links) is larger than the number of orthogonal signals than the available bandwidth may produce. In this situation it is clear that at least some links will cause interference with each other. As the number of terminals and RAPs that maintain connections are increased, the interference will clearly rise and will become the main concern. Fortunately, the varying path loss encountered in the different links in a wireless network will be of benefit, provided these differences can be utilized efficiently. If signals are used in different parts of the service area at a large distance, the mutual interference could be quite low even though the cross correlation between the waveforms is high. In fact, at higher frequencies, where the received signal level decays (almost monotonically) with the distance, the interference power could be reduced to any low level simply by increasing the distance between the receiver/transmitter pairs. If RAPs and terminals are assigned signals with some care, it would even be possible to reuse the same signal several times within the service area with only moderate mutual interference. By reusing the signals, several connections can be established at the same time, and more information can be transmitted. The capacity of the system increases. An apparent paradox is that the higher the propagation loss as function of distance, the lower the interference level and the more often we can reuse the spectrum. This is illustrated in the following example.

Example 4.1

Two mobile radio connections are established along a highway. The transmitters use the same transmitter power and identical waveforms. The modulation and detection schemes are such that a signal-to-interference ratio (SIR) at the receivers has to be at least 10 dB to achieve acceptable communication performance. The link geometry of the system is shown in Figure 4.1. Assume that the propagation models using power law distance dependence and log-normal shadow fading discussed in Section 3.5 are applicable. How large does the distance D_{12} have to be, in order to let receiver M2 reach the SIR requirement with 90% probability if the propagation constant $\alpha = 4$ and 2, respectively.
A. Without shadow fading?
B. With shadow fading, assuming a (log-) standard deviation of 6 dB?

Solution

A. *No shadow fading:*
Schematically, we can describe the system by fig 4.2. Assume that the fading is flat and that the system functions H_{ij} are constants. The received power, may be written as (see Appendix A).

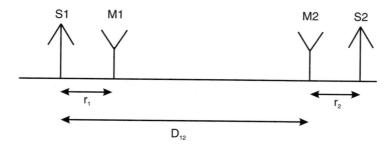

Figure 4.1 Geometry in Example 4.1.

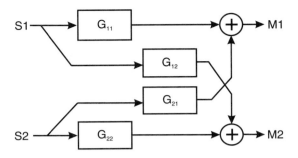

Figure 4.2 System model of the communication system in Example 4.1.

$$P_{rx} = GP_{tx} = \frac{cP_{tx}}{D^a}$$

where P_{tx} denotes the transmitter power and D is the distance between transmitter and receiver. Using this expression, calculate the SIR at receiver M2, denoted by Γ:

$$\Gamma = \frac{\dfrac{cP_2}{r_2^\alpha}}{\dfrac{cP_1}{D_{12}^\alpha}} = \frac{P_2}{P_1}\left(\frac{D_{12}}{r_2}\right)^\alpha \tag{4.1}$$

where P_1, P_2 are the transmitter powers of transmitters S1 and S2 respectively. If these powers are equal, we can use the SIR requirement, $\Gamma > 10$, to yield

$$\left(\frac{D_{12}}{r_2}\right)^\alpha > 10$$

$$D_{12} > r_2 10^{1/\alpha} \approx \begin{cases} 3.2r_2 & \alpha = 2 \\ 1.8r_2 & \alpha = 4 \end{cases}$$

Note the Γ in this case is a deterministic quantity and the required SIR is always achieved as soon as the distance relation above is satisfied.

B. *Shadow fading:*
Again, write

$$P_{rx} = \frac{cGP_{tx}}{D^a}$$

where G is a log-normally distributed random variable with log-mean 0 dB. Compute the SIR in receiver M2 according to

$$\Gamma = \frac{\dfrac{cG_{22}P_2}{r_2^{\alpha}}}{\dfrac{cG_{12}P_1}{D_{12}^{\alpha}}} = \frac{G_{22}}{G_{12}} \frac{P_2}{P_1} \left(\frac{D_{12}}{r_2}\right)^{\alpha}$$

G_{22}, G_{12} are independent r.v:s that in a similar way as above describe the shadow fading in the wanted signal path and the interfering signal path. Again assuming both transmitters use the same power yields:

$$\Gamma = G\left(\frac{D_{12}}{r_2}\right)^{\alpha}$$

One can easily show that G also is a log-normally distributed r.v. with log-standard deviation

$$\sigma_G = \sqrt{2} \cdot 6 \approx 8.5 \text{ dB}$$

Compute the availability as

$$\Pr[\Gamma > 10] = \Pr[10\lg\Gamma > 10]$$

$$= \Pr\left[10\lg G > 10 - 10\,\alpha\lg\left(\frac{D_{12}}{r_2}\right)\right]$$

$$= \Pr\left[\frac{10\lg G}{\sigma_s} > \frac{10 - 10\,\alpha\lg\left(\dfrac{D_{12}}{r_2}\right)}{\sigma_s}\right]$$

$$= Q\left(\frac{10 - 10\,\alpha\lg\left(\dfrac{D_{12}}{r_2}\right)}{\sigma_s}\right) > 0.9$$

since the L.H.S. in the last expression is a Gaussian $N(0,1)$ distributed variable. Evaluating the Q-function gives

$$\frac{10 - 10\,\alpha\,\lg\!\left(\dfrac{D_{12}}{r_2}\right)}{\sigma_s} < -1.28$$

$$D_{12} > r_2\,10^{2.1/\alpha} \approx \begin{cases} 11 r_2 & \alpha = 2 \\ 3.3 r_2 & \alpha = 4 \end{cases}$$

The minimum physical distance between two transmitters using the same waveform, required to achieve a certain link quality, is usually denoted the reuse distance. The inverse of this distance is a rough measure of how many transmitter-receiver pairs that are able to reuse the available bandwidth per unit length.[1] In the example it can be noted that the SIR increases rapidly with the distance D_{12}. In fact, an arbitrary high SIR can be achieved by simply making D_{12} large enough. Furthermore, a very important observation can be made here: the SIR, Γ, does not depend on the absolute distances, but only the ratio between r_2 and D_{12}. This can also be seen in the solution of Example 4.1—the required distance D_{12} is proportional to r_2. This means that our system is scalable, in the sense that we can change the distance scales without affecting the SIR Γ. We have, however, to note that we require the exponent α to be constant over the range of interesting distances. The scaling process cannot be driven to extremes—at very short distances propagation conditions approach (line-of-sight) free-space conditions ($\alpha = 2$). At long distances the radio horizon will affect propagation, which could be interpreted as a very large α. Example 4.1 also illustrates the fact that with a large propagation exponent α, only a very short reuse distance is required. The rapidly increasing propagation loss in the case $\alpha = 4$ enables the spectrum to be reused 2–3 times more often than in the free-space propagation case ($\alpha = 2$). It is also notable that the minimum required SIR strongly affects the reuse distance. As observed in the example, this quantity depends on the modulation and detection schemes employed. Waveforms and detectors more robust to interference will allow for frequent (dense) reuse.

One of the crucial questions in wireless network design is to achieve a system with as high a capacity as possible for a given available bandwidth. This means to design a system that allows for the transmission of most

---

1. Computing the SIR with multiple interferers is of course slightly more complicated. We will return to this problem for the two-dimensional case in the next section.

information or that will support as many terminals/users as possible. Clearly, we will also require a high area availability. In the following section we will now seek to formulate this problem in more precise terms for the class of static channel allocation schemes.

4.2 Static Channel Allocation and Simple Capacity Analysis

When examining the expression of the signal-to-interference ratio in Example 4.1, it becomes obvious that the constraints (4.2) can be satisfied by using short communication distances and by making certain that other interfering users are kept at large distances. This line of reasoning, of course, holds only if the received signal power decays with distance in a reasonably monotonic fashion. Example 4.1 dealt with a simple example with only two transmitter-receiver pairs on a straight line. This problem is now generalized to a large number of access ports that are dispersed in a regular fashion in the service area. With a signal power decaying with distance, it is easy to prove that the terminals should establish connections to the (geometrically) closest access point in order to maximize the signal to interference ratio. The service area may be partitioned into connection regions surrounding each access point. The connection region of an access point is the geometrical region around the access point where the received signal power from that access point is larger than the received power from any other access point and where the signal power is high enough to meet the quality requirement. The connection region is thus always included in the coverage area of the access point as discussed in Section 4.1.

A terminal that is located in a certain coverage region will attempt to connect to the access point of that region. For the sake of simplicity, assume that the terminals are located on a planar surface with a circular symmetric path loss. If we want the same transmission quality in all coverage regions, these regions clearly have to be of uniform shape and size. To completely cover a plane with regions of identical shape and size without overlap is usually called a *tesselation*. Confining the discussion to simple shapes like regular polygons, one may show that only three type of regular polygons have the capability of forming a tesselation: triangles, squares, and hexagons. Since the propagation is decreasing with distance one would expect the coverage regions to be circles which are, however, not capable of providing a tesselation. A reasonable compromise would be the hexagon, which both approximates the circle well and has the tesselation property. Hexagonal shapes are quite frequently found in nature. Besides crystals and snowflakes, the most well-known example is found in bee-hives, where bees build hexagonal cavities for the larvae, called cells.[2] "Cellular radio system" has become the phrase used to describe terminal telephone systems. Figure 4.3 shows the geometry of the (hypothetical) coverage regions of such a hexagonal cellular system. The access port is placed in the center of each cell.

The classical resource allocation scheme that is found in all early mobile telephone systems [1, 2] is the static resource allocation scheme or fixed channel allocation, FCA. In this scheme each access port is assigned a certain fixed number of channels. If the expected number of active terminals is the same in all cells of the network (see Section 4.1), the number of channels assigned to each access port should be the same to provide the same level of service in all parts of the system. Make such an allocation of the C available channels, and divide them in K groups of (approximately) equal size. Each access port (cell) is assigned a group of

$$\eta = \left\lfloor \frac{C}{K} \right\rfloor \qquad (4.6)$$

channels, where $\lfloor x \rfloor$ denotes the integer part of x. The access port has the right to use these channels freely to communicate with its terminals, but

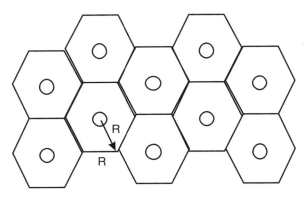

Figure 4.3 Cellular system coverage area layout.

2. In fact hexagonal shape are used by bees because they create a tesselation giving a close fit to neighboring cells, and thus a strong bee-cake. In addition, the hexagon is the tesselating polygon that has the most area compared to its circumference. The latter means that to build a hexagonal cylinder of some given volume requires a minimum of wax.

cannot use any channel from another group. The group size η is a rough measure of the capacity of the system since it indicates the maximum number of simultaneous connections that can be supported in each cell. To maintain a sufficiently high signal-to-interference ratio in the access-port-terminal connections (i.e., larger than γ_0), the channels in a group cannot be reused in a cell that is too close to the first cell. Denote this minimum distance the reuse distance $D(\gamma_0)$. Clearly, many channel groups are used (i.e., large K), and it is obvious that there is no problem in maintaining a large D corresponding to a high γ_0. The penalty paid for such a procedure is, as (4.6) illustrates, that the number of channels that each access port has at its disposal, η, becomes smaller. The capacity of the system will be low. On the other hand, if a low SIR γ_0 can be allowed, D can be too small. In this case, only a few channel groups are necessary and the capacity will be high. Evidently, there is a tradeoff between transmission quality and transmission. This tradeoff is now studied somewhat further.

Assume that the channel groups are numbered $1, 2, \ldots K$. Each cell is thus assigned (labeled with) one of these numbers. A good labeling procedure should maximize the minimum distance between two cells with same label (i.e., maximizing D for a given K). Alternatively, a labeling scheme could be found, that for a given minimum reuse distance D finds the minimum number of required channel groups K. Figure 4.4 shows some channel group assignment, cell plans, for some values of K.

The cell plans shown in Figure 4.4 are examples of fully symmetric cell plans. These cell plans are characterized by the property that the patterns formed by the cells having the same label are identical (just shifted) for all labels. Further, in such a cell plan, each cell has six nearest neighbors, all at

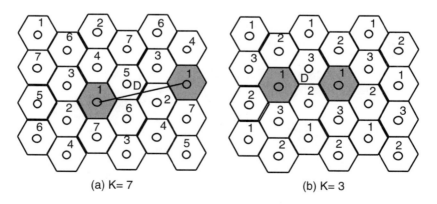

(a) K= 7   (b) K= 3

Figure 4.4 Examples of symmetric hexagonal cell plans.

the same (minimum reuse) distance D. For hexagonal fully symmetric cell plans one can show that the following relationship between K and D holds:

$$\Delta = \frac{D}{R} = \sqrt{3K} \qquad (4.7)$$

R denotes the radius of the cell defined according to Figure 4.3. The quantity Δ is the normalized reuse distance. It is possible to show that there exists fully symmetric cell plans for all integers K that can be written on the form

$$K = (i + j)^2 - ij \qquad i, j = 0, 1, 2, 3 \ldots \qquad (4.8)$$

Possible values are $K = \{1, 3, 4, 7, 9, 12, 13 \ldots\}$ where $K = 1$ corresponds to the trivial case where all channels are used in all cells. For a more detailed discussion and a mathematical analysis of such cell plans, the reader is referred to [3]. In the following example some highly simplifying assumptions are used in the first attempt to assess the capacity of mobile telephony systems based on static resource allocation. The analysis is similar to the one found in [4].

Example 4.2

A mobile telephone system with 100 channels uses a modulation scheme requiring a minimum SIR, γ_0, of at least 20 dB to achieve acceptable link performance. Assume that the propagation loss is only distance dependent and increases with the fourth power of the distance. At most how many channels per cell η can be offered by the system. A symmetric hexagonal cell plan is used with the access ports at the center of the hexagons. All access ports are assumed to use the same transmitter power.

Solution

In a symmetric hexagonal cell plan each cell has exactly 6 co-channel neighbors (using the same channel group) at distance D. Furthermore there are 6 additional co-channel cells at distance $\sqrt{3}D$, 6 at distance $\sqrt{4}D$, 6 at distance $\sqrt{7}D$, ... 6 at distance $\sqrt{K}D$, and so forth, for all K's given by (4.8) (Figure 4.5). Assuming the worst-case situation that the channel is used in all these co-channel cells it is easy to convince oneself that the lowest SIR in the center cell is found when a terminal is on the cell boundary, in one of the

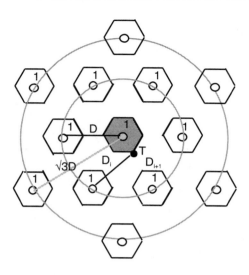

Figure 4.5 Cochannel cell tiers.

corners of the hexagon. If all terminals use the constant transmitter power P, the signal-to-interference ratio at the terminal T can be expressed as

$$
\Gamma_j^d = \frac{PG_{ji_0}}{\displaystyle\sum_{k \in B^{(c_0)}} PG_{ki_0}} = \frac{\dfrac{cP}{R^4}}{\displaystyle\sum_k \dfrac{cP}{D_k^4}} \approx \frac{\dfrac{cP}{R^4}}{6\dfrac{cP}{D^4} + 6\dfrac{cP}{9D^4} + 6\dfrac{cP}{16D^4} \cdots}
$$

$$
= \frac{D^4}{R^4} \frac{1}{6\left(1 + \dfrac{1}{9} + \dfrac{1}{16} + \dfrac{1}{49} \cdots\right)} \approx \frac{1}{7.4}\Delta^4 > 80 \ (19 \text{ dB})
$$

where c is an arbitrary propagation and antenna related constant. In the third step the approximation that all interferers within the same interference tier are at the same distance (e.g., all closest neighbors are at distance D) was made. It can be noted that some of the interferers will actually be closer than the nominal distance, and some will be farther away. Unless very small reuse distances are used, this approximation will not introduce very large errors. A similar analysis can be made in the up-link. We get

$$
D > R(592)^{1/4} \approx 4.95R
$$

Using (4.7) we get

$$K' = \frac{1}{3}\left(\frac{D}{R}\right)^2 \approx 8.11$$

Checking (4.8), there exists a symmetrical cell plan for K = 1, 3, 4, 7, 9, 12. K = 7 will obviously not provide a large enough reuse distance, forcing the use of K = 9 (Δ = 5.2 > 4.95). The quantity η now becomes

$$\eta = \left[\frac{C}{K}\right] = \left[\frac{100}{9}\right] = 11 \text{ channels/cell}$$

A common approximation that is often used for quick approximate calculations is to neglect all interferers but the first tier of closest neighbors. In the example above, this would yield the same end result. (As an exercise, show this to be true.) The single tier approximation should, however, be used with some care, in particular, if the propagation exponent is less than four. In general the SIR can be written as

$$\Gamma_j^d \approx \Delta^\alpha \frac{1}{6} \frac{1}{\sum\limits_{k \in K} k^{-a/2}} \tag{4.9}$$

where K denotes the (infinite) set of numbers satisfying the relation (4.8). One can show that the sum in the denominator does not converge if α is less or equal to two (e.g., free space propagation). This phenomenon, yielding infinite interference power, is usually referred to as Olbers paradox,[3] which is well known in astronomy. In practice this is not a real problem, since all networks are finite. In addition, other propagation phenomena, such as the radio horizon, limit the interference power. It is clear, however, that for moderate and low α's, great care has to be exercised not to neglect distant interferers.

---

3. Olbers paradox: Why is the sky dark at night? If all stars were uniformly distributed over the volume of the universe, the expected (average) number of stars at some given distance r from earth, would increase as r^2. The radiation power received by an observer on the earth from these stars would be proportional to r^{-2}. Summing all the power from every star over an infinite universe would yield an infinite total received power.

4.3 Traffic-Based Capacity Analysis

The number of available channels in every access port is a rough but useful capacity measure. It describes the capability to serve users from an operator perspective, but it does not reflect the user satisfaction. For this purpose start with introducing the assignment failure rate (3.4) as the user performance constraint. Now, proceed by determining the capacity as given by (3.5) of the (hypothetical) static channel allocation. In this allocation scheme, the channels in each cell can be freely allocated to any terminal in that cell, independently of what is going on in other cells. The worst case interference assumption will guarantee that the SIR condition is fulfilled for any of these channel allocations. Therefore, consider the allocation of a channel in one particular cell. Assuming that the number of terminals in this cell is M_c, the number of assignment failures can be computed as

$$Z = \max(0, M_c - \eta)$$

Using the definition (3.4), the assignment failure rate can be found

$$\nu_p = \frac{E[2]}{E[M]} = \frac{E[\max(0, M - \eta)]}{\omega A_c} = \sum_{\kappa=\eta}^{\infty} (k - \eta) \frac{(\omega A_c)^{k-1}}{k!} e^{-\omega A_c}$$

(4.10)

where the fact that M_c is Poisson distributed with expected value ωA_c has been used. Further

$$A_c = \frac{3\sqrt{3}}{2} R^2$$

(4.11)

denotes the area of the cell. Figure 4.6 shows the assignment failure rate ν_p (4.10) some typical values of η as function of the *relative traffic load* ϖ_η, defined as

$$\varpi_\eta = \frac{\omega A_c}{\eta}$$

(4.12a)

An alternative measure of the traffic load is the expected number of terminals per cell per the total number of channels in the entire system, C,

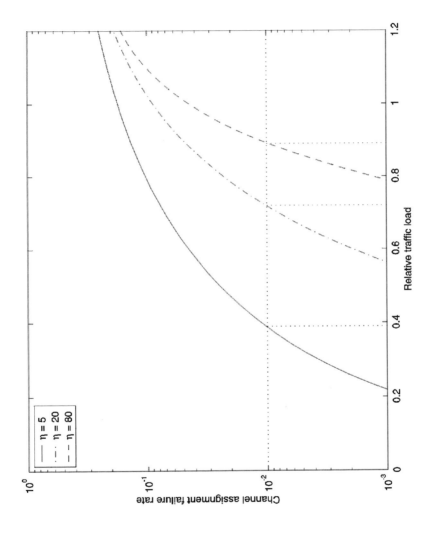

Figure 4.6 Channel assignment failure rate as function of relative traffic load ϖ_η for $\eta = 80$, 20 and 5 channels/cell.

$$\varpi_c = \frac{\omega A_c}{C} \approx \frac{\varpi_\eta}{K} \qquad (4.12b)$$

As evident from Figure 4.6, the assignment failure rate is increasing function in ω and decreasing in η. Less obvious is the fact that the assignment failure rate is decreasing in η when we increase ω with the same amount, i.e., keeping ϖ_η constant. A well-known result from traffic theory is that a system with many channels is more efficient than a system with fewer channels ("the bigger the better," trunking-gain). This result is illustrated with the following example:

Example 4.3a

A cellular telephone system designed for 9 channel groups has a cell radius of 1 km. What is the capacity of the system (measured in calls/km²) if we allow an assignment failure rate of at most 1%) and the system has C = 720, 180, or 45 channel pairs at its disposal?

Solution:

Determine the area of the cell:

$$A_c = \frac{3\sqrt{3}}{2} 1^2 \approx 2.6 \text{ km}^2 \qquad (4.13)$$

The number of available channel pairs/cell is given by (K = 9)

$$\eta = \left\lceil \frac{C}{K} \right\rceil = 80, 20 \text{ and } 5 \text{ channel pairs/cell} \qquad (4.14)$$

From Figure 4.6 we find

$$\varpi_\eta = \frac{\omega A_c}{\eta} = 0.89, 0.72, \text{ resp } 0.39 \qquad (4.15)$$

for η = 80, 20 and 5 respectively at 1% assignment failure rate. Combining (*1), (*2) and (*3) we get

$$\omega = \frac{\eta \varpi_\eta}{A_c} = 27, 5.5 \text{ and } 0.75 \text{ calls/km}^2$$

Note that the number of channels/cell η is increased by a factor of four between the different cases. The capacity increases by a factor of 7 and 5 respectively due to the trunking gain.

Notable in the example is that the capacity is primarily dependent of η, the number of channels per cell and the cell area, A_c. Rewriting the last expression in the example yields

$$\omega = \frac{\eta\varpi_\eta}{A_c} = \frac{C\varpi_\eta}{KA_c} \qquad (4.16)$$

Using the result in (4.9) and (4.7), the following approximation can be made

$$K \approx \frac{\Delta^2}{3} \approx c(\alpha)\gamma_0^{2/\alpha}$$

where $c(\alpha)$ is a constant and γ_0 is the minimum required SIR. Inserting this result in to (4.16) we get the following approximate result:

$$\omega \approx c'(\alpha)\frac{C\varpi_\eta}{\gamma_0^{2/\alpha}A_c}\omega \qquad (4.17)$$

As we can see from expression (4.17), there are in principle, three ways to increase the capacity of an wireless communication system:

- Increasing C: Fairly obvious—the drawback is the increased bandwidth (increasing proportionally to C) Since (new) spectrum resources may be hard to come by this could be difficult.

- Decreasing γ_0: Employing more interference resistant modulation/ detection and coding schemes can do this. Since this results in decreased K, and thus increased η, the gain is twofold since the factor ϖ_η is also increased (see Example 4.3a). Unfortunately there is usually a bandwidth penalty (i.e. a decrease in C and thus possibly also η), but net gains in capacity can usually be achieved.

- Decreasing A_c: By decreasing the size of the cells there is in fact no real limit to how large the capacity may become. Very small A_c will improve propagation (i.e., to line-of-sight conditions $\alpha = 2$) which will reduce the constant c' somewhat. The penalty is the increasing number of cells (access ports) per area unit. Since the number of access ports is inversely proportional to A_c, the capacity will be roughly proportional to the total number of access ports in the system.

Another observation is that poorer propagation conditions, that is, a decreasing α in fact improves the capacity of the system.

Systems using static resource allocation are also quite easy to analyze using the traditional traffic model for telephony. If the assumption that the mobility of the terminals is rather limited (i.e. the terminals generally stay within a single cell under the duration of each call) is made, the call handling in each cell can be modeled as independent M/M/η—(Erlangian) blocking systems. This type of queuing system is characterized by calls arriving according to a Poisson process (independent inter-arrival times with exponential distribution). The arriving calls are "served" by "servers" (the η available channels per cell). Calls are blocked and disappear if all channels are found to be busy. The classical performance measure here is the blocking probability, that is, the probability that a newly arriving call finds all channels busy and is denied service. It is well known that the blocking probability is given by

$$E_\eta(\rho_c) = \frac{\dfrac{\rho_c^\eta}{\eta!}}{\displaystyle\sum_{k=0}^{\eta} \frac{\rho_c^k}{k!}} \qquad (4.18a)$$

where the relative traffic intensity per cell is defined as

$$\rho_c = \frac{\lambda A_c}{\mu A} = \rho \frac{A_c}{A} \qquad (4.18b)$$

The expression (4.18a) is also know as the Erlang-B formula. Figure 4.7 shows the blocking probability as function of ρ_c/η, the relative traffic load in Erlangs/channel, for some different η. We clearly note the similary with the assignment failure probability results in Figure 4.6 Below we now apply our new results to our previous example (4.3). As can be seen from the example, even the numerical results obtained are almost identical.

Example 4.3b

A mobile telephone system with slowly moving terminals uses a cell plan with 9 channel groups and a cell radius of 1 km. What is the capacity of

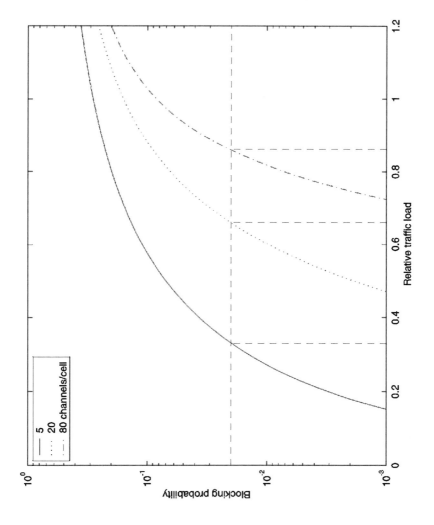

Figure 4.7 Blocking probability as function of relative traffic intensity per channel and cell.

the system (measured in Erlang/km^2 at a maximum blocking probability of 2%) if C = 720, 180 or 45 channel pairs are available?

Solution:

As in example 4.3a, determine the area of the cell and the number of available channel per cell η. By means of the graphs in Figure 4.7 the relative traffic load corresponding to 2% blocking probability is determined for η = 80, 20 and 5 respectively:

$$\frac{\rho_c}{\eta} = 0.87, \ 0.67, \text{ and } 0.33$$

$$\rho_A \frac{\rho_c}{A_c} = \frac{\rho_c}{\eta} \frac{\eta}{A_c} \approx 27, \ 5 \text{ and } 0.65 \text{ Erlang/km}^2$$

4.4 Outage-Based Capacity Analysis

4.4.1 Interference Outage

The "worst-case" design method used in the previous section gives a rather coarse picture of the interference situation in a cellular communication system. In the analysis, so far a very simple propagation model (example 4.2) was used. As will become evident in the following, this model tends to produce rather optimistic capacity results. On the other hand, in the design of a cell-plan, the rather pessimistic assumption was made, that a certain SIR was to be achieved in all locations in the cell under all interference conditions. This led to the requirement that the prescribed SIR should be reached at the corner of the cell when all potential users of a channel were actually using it. Provided it is possible to design such an interference-free cell-plan, all assignment failures will be caused by the fact that the number of channels in a cell is not sufficient.

In real cellular systems, the received signal levels are not smoothly distance dependent. On the contrary, due to shadow fading the received signal may fluctuate considerably. The straightforward approach to handling the signal variation is to include a fade margin on top of the minimum required SIR. To guarantee that practically *all* terminals in this way obtain an adequate SIR would require impractically large reuse distances and thus uninterestingly low capacities. Any real cellular system will thus trade off a small fraction of terminals not satisfying the SIR-requirement for a higher capacity. On the other hand, assuming that all channels are used is very pessimistic. As is evident from the previous section it is known that this would correspond to a very high traffic load at which newly arriving calls always would find all channels busy. In this section we will therefore refine the analysis somewhat in order to study the two effects mentioned above more carefully.

For this purpose, we start by introducing the stochastic variable Q, denoting the number of terminals that have been assigned a channel, but that cannot obtain an adequate SIR. From a users perspective, having received a channel does no good if the channel turns out to be useless. Fast moving terminals may have some hope that they move into a more favorable situation quickly, as opposed to slowly moving terminal. In fact, from the user's point of view, in the latter case it may not even be possible to distinguish between a conventional assignment failure and a poor channel. Communication is not possible in both these cases. For this latter case, the performance measure can now be generalized, and the rate at which the system fails to assign useful channels is termed the (total) assignment failure rate defined as:

$$\nu = \frac{E[Z]}{E[M_c]} = \frac{E[\max(0, M_c - \eta)] + E[Q]}{\omega A_c}$$

$$= \frac{E[\max(0, M_c - \eta)]}{\omega A_c} + \chi = \nu_p + \chi \qquad (4.19)$$

where as in the previous section M_c, Z_c have been used to denote the active and not properly assigned mobiles in each cell, respectively. Furthermore, the following interference rate was introduced

$$\chi = \frac{E[Q]}{\omega A_c} \qquad (4.20)$$

Already at moderate traffic loads ω, the interference rate will closely approximate the outage probability

$$\chi \approx \Pr\{\Gamma < \gamma_0\}$$

The primitive assignment failure rate ν_p is therefore nothing more than the assignment failure rate for a system design for cochannel interference-free conditions (see "worst case design" in previous section). The interference rate in a hexagonal, symmetric cell plan is now determined. As in Example 4.2, the SIR in the uplink of a mobile at distance r is studied

$$\Gamma^u = \frac{P_j G_{i_0 j}}{\displaystyle\sum_{k \in M^{(c0)}} P_k G_{i_0 k}} = \frac{P_j G_{i_0 j}}{\displaystyle\sum_{k=1}^{M} P_k X_k G_{i_0 k}} \qquad (4.21)$$

where the activity variable X_k is defined as

$$X_k = \begin{cases} 1 & \text{mobile in cell } k \text{ is active} \\ 0 & \text{otherwise} \end{cases}$$

This quantity is a sum of a random number of independent log-normally distributed random variables. As the channel assignment in the different cells can be assumed to be independent, the number of terms in this sum will be binomially distributed with probability

$$\begin{aligned} q = \Pr\{X_k = 1\} &= \frac{E \min[(M_c, \eta)]}{\eta} \\ &= \sum_{k=0}^{\eta} \frac{k}{\eta} \frac{(\omega A_c)^k}{k!} e^{-\omega A_c} + \sum_{k=\eta+1}^{\infty} \frac{(\omega A_c)^k}{k!} e^{-\omega A_c} \qquad (4.22) \\ &= 1 - \sum_{k=0}^{\eta} \left(1 - \frac{k}{\eta}\right) \frac{(\omega A_c)^k}{k!} e^{-\omega A_c} \le \varpi_c \end{aligned}$$

The RHS in the last expression in (4.22), proves to be a good approximation for q, except when q is close to unity (saturated system). q is denoted as the *activity factor*.

4.4.2 Analytical Approach

Assuming that the powers are the same for all transmitters and that the receiver noise can be neglected, (4.18) can be rewritten as

$$\Gamma^u(r) = \frac{P G_{i_0 j}}{\displaystyle\sum_{k=1}^{M} P X_k G_{i_0 k} + N} \approx \frac{G_{i_0 j}}{\displaystyle\sum_{k=1}^{M} X_k G_{i_0 k}} = \frac{G_0 \left(\dfrac{1}{r}\right)^\alpha}{\displaystyle\sum_{k=1}^{M} X_k G_k \left(\dfrac{1}{D_k}\right)^\alpha} \qquad (4.23)$$

The G_k are assumed to be independent log-normally distributed random variables with log-expectation (= median) 0 dB. Assuming that the interference will be dominated by the first tier of interferers at (approximately) the reuse distance D, rewrite the expression as

$$\Gamma^u(r) = \frac{G_0\left(\frac{1}{r}\right)^\alpha}{\displaystyle\sum_{k=1}^{6} X_k G_k \left(\frac{1}{D}\right)^\alpha} \approx \frac{G_0}{\displaystyle\sum_{k=1}^{6} X_k G_k} \left(\frac{D}{r}\right)^\alpha = \frac{G_0}{G_I}\left(\frac{D}{r}\right)^\alpha$$

The composite fading of the interfering signals is described by the random variable G_I. The PDF of G_I, here denoted $F_I(\gamma)$ is illustrated in Figure 4.8 for σ_s = 6 dB. Note that at high loads, the distribution of G_I is closely approximated by a log-normal distribution (in this case with the parameters ($m_I \approx$ 10 dB, $\sigma_I \approx$ 3 dB) [SWY82, BEA95] The log-normal approximation does also work reasonably well for lower loads. As can be seen from Figure 4.8, the poorest accuracy is found at the tail of the distribution. This is due to the fact that the event of having no interferer, that is, zero interference power, has a positive probability. In fact, expressed in dB, we have

$$F_I(-\infty) = \Pr[\text{no active interferer}] = (1 - q)^{-6}$$

Considering the capacity analysis, systems with a rather high load will be our main concern. This makes the log-normal approximation for G_I feasible. In this case rewrite (4.21) according to

$$\Gamma(r) \approx \frac{G_0}{G_I}\left(\frac{D}{r}\right)^\alpha = G_{\text{tot}} \delta^\alpha$$

where $\delta = D/r$ was introduced $G_{\text{tot}} = G_0/G_I$ is also a log-normally distributed variable with parameters

$$m_{\text{tot}} = - m_I$$

$$\sigma_{\text{tot}} = \sqrt{\sigma_I^2 + \sigma_s^2}$$

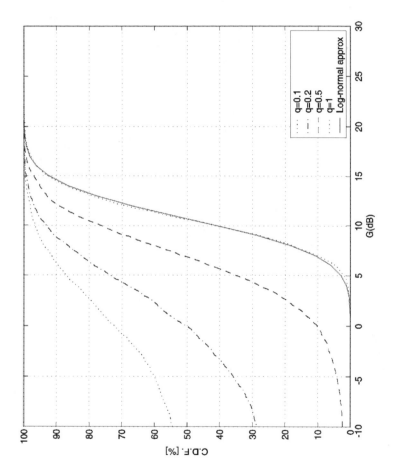

Figure 4.8 Probability Distribution Function (PDF) of a sum a binomially distributed number of log-normally distributed variables G_I for different activity probabilities q. The log-standard deviation of the components $\sigma = 6$ dB. Dashed line indicates log-normal variable where we matched parameters ($m_I \approx 10$ dB, $\sigma_I \approx 3$ dB).

One can now write the interference rate ($\approx$ outage probability) at a mobile at distance r as

$$\chi(r) \approx \Pr\{\Gamma \leq \gamma_0 \mid r\} = \Pr\{G_{\text{tot}}\delta^\alpha \leq \gamma_0\}$$

$$= \Pr\left\{G_{\text{tot}} \leq \frac{\gamma_0}{\delta^\alpha}\right\} = \Pr\{10\lg G_{\text{tot}} \leq (\gamma_0)_{\text{dB}} - 10\alpha\lg\delta\} \quad (4.24)$$

$$= \Phi\left(\frac{(g_0)_{\text{dB}} - 10\alpha\lg\delta + m_I}{\sigma_{\text{tot}}}\right)$$

where Φ is the PDF of a Gaussian $N(0,1)$ distributed random variable. The interference rate in the up-link of a randomly chosen terminal can now be computed as

$$\chi \approx \int\limits_{r=0}^{R} \Phi\left(\frac{(\gamma_0)_{\text{dB}} - 40\lg\dfrac{D}{r} + m_I}{\sigma_{\text{tot}}}\right) p(r)\,dr \quad (4.25)$$

$$= \int\limits_{0}^{1} \Phi\left(\frac{(\gamma_0)_{\text{dB}} + 40\lg(x) - 20\lg(3K) + m_I}{\sigma_{\text{tot}}}\right) 2x\,dx$$

where in the final step the variable substitution ($x = r/R$) has been made and (4.6) has been used. Finally, approximate the hexagonal cell by a circle to derive a simple approximation of the density function of the distance of the terminal from the origin (see Exercise 4.13). There is no closed form expression for the function

$$\Psi(a, b) = \int\limits_{0}^{1} \Phi\left(\frac{40\lg(x) - a}{b}\right) 2x\,dx \quad (4.26)$$

Figure 4.9 shows the behavior of this function. These results are applied in Example 4.4.

Example 4.4

In the mobile telephone system in Example 8.2 ($C = 100$ channels, $\gamma_0 = 20$ dB), assume the path losses to be log-normally distributed

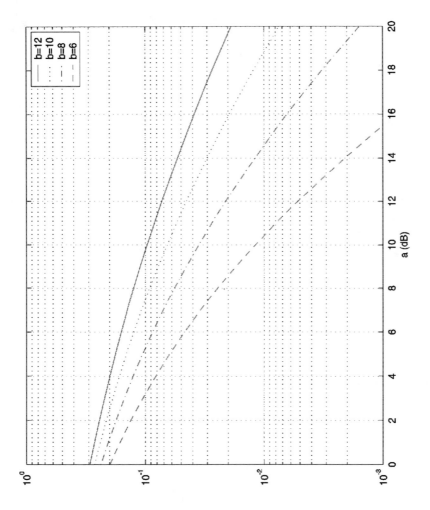

Figure 4.9 The function $\Psi(a; b)$ for $b = 6$, 8, 10 and 12 (dB).

(σ_s = 6 dB) and independent with a log-average increasing proportional to $40 \log(r)$.

    a. Determine the interference rate χ at high traffic loads.

    b. Determine the number of channel groups required to achieve an interference rate of at most 5% for the high load case.

Solution:

At high traffic loads the approximation $q \approx 1$ is made which from Table 4.1 immediately gives

$$m_I \approx 10 \text{ dB}, \qquad \sigma_I \approx 3 \text{ dB}$$

$$\sigma_{\text{tot}} = \sqrt{\sigma_I^2 + \sigma_s^2} = \sqrt{3^2 + 6^2} \approx 6.7 \text{ dB}$$

    a. We identify the parameters in eq. (4.23) using $K = 9$

$$a = 20 \lg(3K) - (\gamma_0)_{\text{dB}} - m_I = 28.6 - 20 - 10 = -1.4 \text{ dB} \quad (*)$$
$$b = 6.7$$

Using the graph in Figure 4.9 (and making a simple interpolation) yields the approximate result

$$\chi = 20\%$$

    b. Again using fig 4.9 to interpolate between the curves for $b = 6$ and $b = 8$ in order to find the a satisfying the relation

$$\Psi(a; 6.7) = 0.05$$

Table 4.1
Parameters of Log-normal Approximation $\alpha = 4$

q	σ_s = 6 dB		σ_s = 8 dB		σ_s = 10 dB	
	m_I	σ_I	m_I	σ_I	m_I	σ_I
0.5	7	4	8	6	9	7
1.0	10	3	12	4	14	5.5

yielding

$$a \approx 6.5 \text{ dB}$$

Inserting this result into (*) above, we get

$$20 \lg(3K) > 6.5 + (\gamma_0)_{\text{dB}} + m_I = 36.5$$
$$K > 22.3$$

Chose (using (4.7)) $K = 25$.

Note that the design in Example 4.2, which was made neglecting fading, yields a far from negligible interference rate. In fact, roughly 20% of the channel will turn out to be useless. The simple models thus yield overly optimistic estimates of the system capacity. Adding a fade margin by increasing the reuse distance on the other yields a much lower capacity compared to the original estimate in (4.2) (not even considering the trunking loss).

The results from the interference analysis above and the assignment failure rates can now be combined to present a more complete capacity analysis of a wireless network in shadow fading. Consider the following example:

Example 4.5

Design a mobile telephone system using $C = 400$ channel pairs. The modulation and coding schemes employed require a minimum SIR of 13 dB. The propagation losses are assumed to be uniform and independent in all directions, log-normally distributed ($\sigma_s = 6$ dB) with a log-average increasing proportional to $40 \log(r)$. Determined the minimum number of channel groups and estimate the capacity (in calls/cell) if the total assignment failure rate has to be kept below 2%?

Solution:

The purpose is to determine the system capacity, i.e., the behavior of the system at high loads. Initially we will assume that the activity probability is high and make the approximation $q \approx 1$. We now compute χ for some different values of K in the same manner as in example 8.4. Locking the total assignment failure rate ν to the 2% level will now immediately yield the primitive assignment failure ν_p as

$$\nu_p = \nu - \chi$$

In the same manner as in example 8.3a the capacity can now be computed. The following small table can be generated:

Step 1
"Full Load" ($q = 1$)

K	η	χ	ν_0	ν	$\varpi\eta$
16	25	2%	0	2%	0
19	21	1%	1%	2%	15
21	19	1.3%	0.7%	2%	14
25	16	0.4%	1.6%	2%	13

This approximate analysis indicates that there is a shallow maximum around $K = 19$ yield a capacity of about 15 mobiles/cell. Applying this load to the system will clearly render our "full load" approximation pessimistic, with respect to the interference power. In fact we have $q \approx \varpi_c \approx 0.7$. For this load , use table 4.1 to find

$$m_I \approx 8.5 \text{ dB} \qquad \sigma_I \approx 3.6 \text{ dB}$$

$$\sigma_{\text{tot}} = \sqrt{\sigma_I^2 + \sigma_s^2} = \sqrt{3.6^2 + 6^2} \approx 7.0 \text{ dB}$$

Reiterating the previous procedure for this new interference situation yields the following table:

Step 2
"True Load" ($q \approx \varpi_c \approx 0.7$)

K	η	χ	ν_0	ν	$\varpi\eta$
13	30	>2%	—	—	—
16	25	1.2%	0.8%	2%	17
19	21	0.6%	1.4%	2%	16

From this table $K = 16$ is found to be the best choice yielding a capacity of about 17 active mobiles/cell. (This load also corresponds to $q \approx \varpi_c \approx 0.7$. Thus the iteration process can be stopped and the current result can be considered as reasonable.)

In the example above, again note the trade-off between capacity and quality discussed in the previous section. A high primitive assignment failure rate can be traded off against high interference rates. Choosing a small K creates many channels, such that there will be one available almost all the time. However, due to the low reuse distance, the average quality of these channels will be low, resulting in a high interference probability. On the other hand, a large K will create a small number of channels that are usable with high probability.

Furthermore note that the results obtained by the worst case analysis without fading are certainly overly optimistic (the corresponding result would yield $K = 4$ and a capacity close to 90 mobiles/cell) even though only the six closest interferers were considered.

4.4.3 Simulation Approach

The approximations given in the previous chapter are in some cases convenient, but they have their limitations. In particular, in a system with a rather small number of available channels C, the load factor q is low ($q < 0.5$) when reasonable assignment failure rates are required. As can be seen in Figure 4.8, the Gaussian (log-normal) approximation of the sum G_I in the denominator of (4.20) breaks down. A similar problem occurs if the propagation constant α is smaller than 4. In this case it may not be possible to neglect the potential interferers outside the first tier of cochannel neighbors, in particular, if the log-standard deviation of the shadow fading is large. Taking more interferers with different average powers into account definitely complicates the approximation.

A third complication is the fact that in most systems, the access port selection will be such that the instantaneously strongest access port is selected. Due to the shadow fading, this is not necessarily the access port in the center of the cell the terminal is located in (i.e., the geographically closest one). Such a far away terminal is likely to create more damaging interference (or will be more susceptible to interference) than the terminals in the original model, which are confined to the cell with the serving access port. Figure 4.10 illustrates this by showing a typical realization of mobile terminal locations. All terminals are connected to the access port with the instantaneously largest path gain. In those cases when the terminal connects to an access port outside "its own" cell, a small dash indicates the direction to the serving access port.

Although there are analytical approximation techniques that take all these factors into account, it is in many cases more convenient to use computer

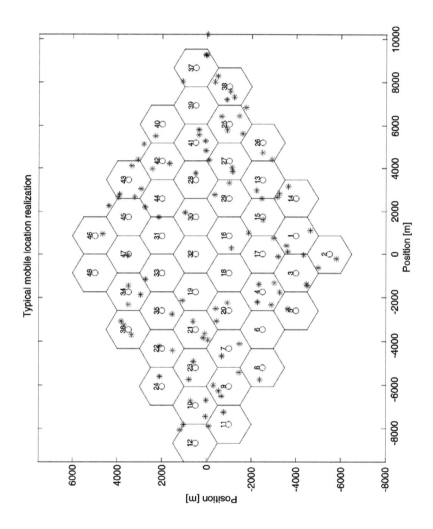

Figure 4.10 Typical mobile location realization (RUNE).

simulations to estimate the desired quantities of interest. Ultimately, the RUNE tool, which is described in Appendix B and C, will be to make a more refined capacity analysis. RUNE is a MATLAB™ toolbox containing datastructures suitable to describe signal and interference levels, powers in cellular systems and functions to manipulate them.

Figure 4.11 shows an example, the CDF in the (uplink) of the C/I in a cellular system of the type we discussed in the previous section for various reuse factors $K = 3,7, \ldots 16$ in a realistic load situation. Comparing the results in this graph with the Example 4.5, where the optimum solution required an outage probability of between 1–2% at $K = 16$ for a 13 dB C/I value, shows that the analytical scheme in this case seems to be working quite reasonable. Other results that can be derived by the simulation model is dependent on the received signal power and/or the external (thermal) noise level.

In large cells (macro-cells) the received power will tend to be rather low at the cell boundaries and the signal-to-noise ratio (C/N) will also be low. The opposite will hold true in cells with a small radius. In large cells the performance will tend to be limited by the noise (or range), whereas in smaller cells, the interference will dominate as discussed in Chapter 3. To quantify this, define $\overline{\Gamma}(R)$ as the median C/N at distance R from the base station.

$$\overline{\Gamma}(R) = \text{median}\left\{ \frac{PG_{i_0j}}{\sum_{k=1}^{M} PX_k G_{i_0k} + N} \right\} \approx \text{median}\left\{ \frac{G_0 P\left(\frac{1}{R}\right)^{\alpha}}{N} \right\} = \frac{P}{NR^{\alpha}}$$

(4.27)

A preference for the median instead of the mean is obvious from the last step in the equation, since for G_0, a log-normal variable, the median can be computed:

$$\Pr[G_0 \leq g] = \Pr\lfloor 10^X \leq g \rfloor = \Pr[X \leq \lg(g)] = 0.5$$

Noting that X is a zero-mean (and zero median!) Gaussian variable it can be concluded that $\overline{X} = 0 \Rightarrow \overline{G}_0 = 1$. Figure 4.12 shows the C/I distributions for two situations, a noise limited case ($\overline{\Gamma} = 20$ dB) and a purely interference limited case ($\overline{\Gamma} = \infty$) corresponding to our previous noise-less analysis. For $K = 3$, the difference between the dotted and the solid curves

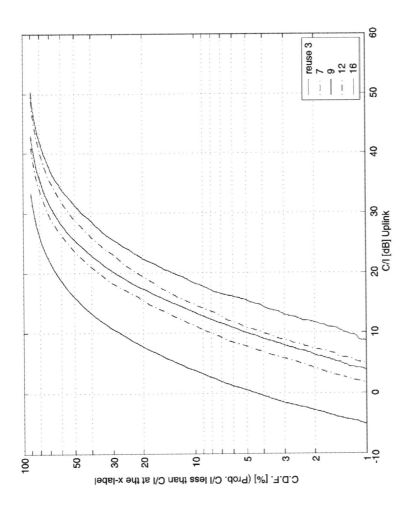

Figure 4.11 *C/I* distribution: reuse factors (from left) K = 3, 7, 9, 12, 16. Omnidirectional antennas. Load q = 0.5, α = 4, log-normal fading: σ = 8 dB.

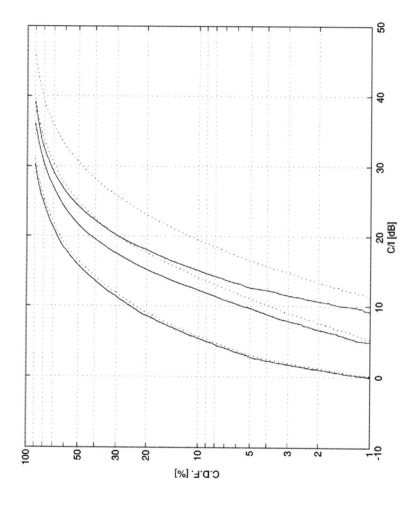

Figure 4.12 *C/I* distribution: reuse factors (from left) K = 3, 7, 12; Solid lines: large cell radius (noise limited system); $\overline{\Gamma}$ = 20 dB; Dotted lines: Small cell radius (Noiseless system), $\overline{\Gamma}$ = ∞, Omnidirectional antennas. Load q = 0.8. α = 4, log-normal fading: σ = 8 dB.

is not that large, since the interference tends to dominate in this case. However, as the reuse factor is increased, the noise-limited system exhibits much lower performance. Due to the dominance of the noise for large reuse factors, going from $K = 7$ to $K = 12$ yields hardly any improvement in the C/I distribution.

The impact of the load factor, q, is illustrated in Figure 4.13. Again, the same interference limited and noise limited scenarios as in Figure 4.12 have been used. As may be expected, for the low reuse factors (e.g., $K = 1, 3$), the traffic load has a strong impact on the outage probability since here the cochannel interference is a strongly dominating source of disturbance. For higher reuse factors, the dependence on the load decreases and outage probability curves become more and more flat.

Now, take a look at the combined effect of the outage and the assignment failure rate. The outage probability, the primitive assignment failure rate (blocking probability) and the total assignment failure rate (here denoted the GOS) are plotted for two different reuse factors ($K = 3, 7$) as function of the load per cell per total channel in the system, ρ_c, in Figure 4.14. From the definition of ϖ_c in (4.12b) we can derive

$$q \approx \varpi_\eta = K\varpi_c$$

As may be expected, in the system with the higher reuse factor, $K = 7$, the growing interference caused by the increasing traffic load is not a serious problem. Instead, the blocking probability is the main concern here. When the load per channel, ϖ_c gets close to $1/K$, the load factor q approaches 1, with high blocking probabilities as consequence. As can be seen in the graph, this latter effect is completely dominating and the GOS-curve, in principle follows the blocking probability curve. This system is subject to a shortage of channels and thus blocking is limited. For the low reuse-factor case, $K = 3$, the situation is reversed. Here blocking becomes a problem when ϖ_c gets close to 1/3, but far before that the outage probability has risen to dominate the scene. For practical (low) values, the GOS is dominated by the outage probability, that is, by the link quality of the channel and we have an interference limited system. Again this illustrates the choice between creating many channels with poor link quality (small K) or limiting the design to few channel with good link quality (large K).

Looking again at Figure 4.14. The GOS curves intersect at GOS $\approx 7\%$ for which the capacity $\varpi_c \approx 0.1$ mobiles/cell/channel, that is, $\omega = C\varpi_c = 10$ mobiles/cell can be obtained. For the $K = 7$ systems this would correspond to a $q = 0.7$, an assignment failure probability of about

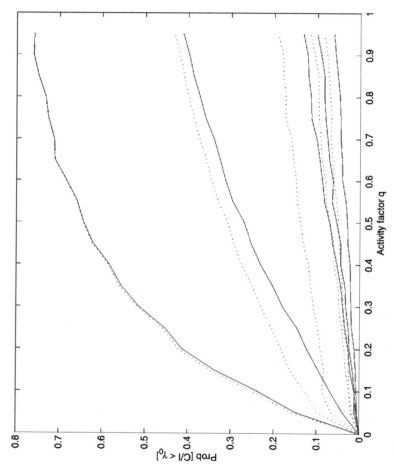

Figure 4.13 Outage probability as function of load factor q: Reuse factors (from top-left): $K = 1, 3, 7, 9, 12$. Dotted lines: Large cell radius (Noise limited system), $\overline{\Gamma} = 20$ dB, Solid lines: Small cell radius (Noise-less system), $\overline{\Gamma} = \infty$, Omnidirectional antennas. $\alpha = 4$, log-normal fading: $\sigma = 8$ dB.

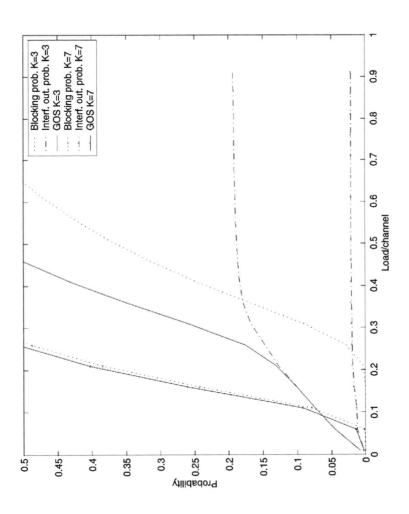

Figure 4.14 Outage (Interference, dotted), Primitive assignment failure (blocking) probabilities (dashed) and the total assignment failure rate (GOS, solid) as function of load/channel in system, ϖ_c. $K = 3$ and 7. $C = 100$ channels.

1% and 6% outage probability. More interesting is to note that the same GOS can be obtained by the $K = 3$ system. Here, $q = 0.3$ and the assignment failure probability can be neglected. This is called a fractionally loaded system. In a blocking limited system, the interference level is not a problem, since there will be a shortage of channels which will limit the number of mobiles getting access to the system and causing excessive interference. In the $K = 3$ system operating at a fractional load, however, there will be plenty of channels available to newly arriving calls. Therefore the classical blocking mechanism will not be sufficient to control the interference in the system to maintain the prescribed GOS-level. Some interference power based admission control system will have to be introduced, preventing too many users to enter the system. Admission control schemes will be discussed further in Chapters 8 and 9. To round this section off, an example is provided to show how the simulation tools can be used for a simple design, similar to the one derived in Example 4.5.

Example 4.6

Design a mobile telephone system using $C = 100$ channel pairs. The modulation & coding requires a C/I of 13 dB to provide sufficient link quality. Define the GOS as

$$GOS = \chi + P_{bl}$$

Determine the capacity of the system in Erlangs/cell if a minimum GOS of 15% is to be achieved.

Solution:

In the same manner as in Example 4.6, start by evaluating the activity factor q is and make the approximation $q \approx 1$. We now compute χ for some different values of K in the same manner as by looking at the C/I distributions derived from our RUNE model as in Figure 4.15. The following small table is the result:

K	χ
3	55%
7	27%
9	20%
12	10%

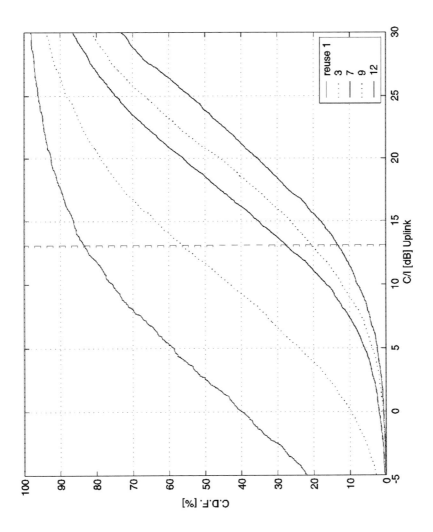

Figure 4.15 Illustration for Example 4.6.

Locking the GOS to a 15% level will yield the blocking probabilities as

$$P_{bl} = 15 - \chi \ (\%)$$

Step 1
"Full Load" ($q = 1$)

K	η	χ	P_{bl}	GOS	ρ
12	8	12%	3%	15%	3
9	11	20%	***	***	0
7	14	27%	***	***	0

This approximate analysis indicates that there is a shallow maximum around $K = 9$ with a load of about 5 Erlang/cell. The activity factor q for this design would be $q \approx 5/11 \approx 0.5$. Deriving new C/I graphs for this load and reading them in the same way as above will yield the next table:

Step 2
"True Load" ($q \approx \varpi_c \approx 0.4$)

K	η	χ	P_{bl}	GOS	ρ
12	8	8%	7%	15%	5
9	11	10%	5%	15%	7
7	14	15%	***	***	0

From this table $K = 9$ is still found to be the best choice, yielding a capacity of about 7 Erlangs/cell.

In the analysis, the assumption has been made that the lognormal shadow fading variables G_i in (4.20) all are uncorrelated. This is a reasonable assumption in the uplink where the interfering signals reach the RAP from mobiles in all directions (due to the uniform location of all mobiles). In the downlink direction, the situation is, however, somewhat different. The mobile will receive signals from a few very distinct directions. In addition, if a mobile terminal moves into some poor location (indoor or behind some large obstacle) many or all RAP will be shadowed more or less simultaneously. On the other hand if the terminal moves to some high location with good reception, it is likely that many, or even all, RAPs will be received with high received signal levels. In the RUNE model, a very simple model for deriving the G_ks in (4.20) is used. First generate independent Gaussian variables Y_M, Y_{1B}, Y_{2B} ... and then compute

$$Y_k = \sqrt{\rho}\,Y_M + (\sqrt{1-\rho}\,)\,Y_{kB}$$

$$G_k = 10^{\,Y_k/10}$$

Note, that for $\rho = 1$, all shadow fading variables are the same for all RAPs. The left graph in Figure 4.16 illustrates this special case. The scatter plot shows the downlink interference levels versus the downlink signal levels in one RUNE simulation. Note the strong correlation between the signal levels. The remaining variations in this case are now only dependent on the variations in the distance dependent path loss due to the random locations of the terminals. Figure 4.20 finally shows the CDF of the downlink C/I for some different values of the correlation coefficient ρ. As can be seen, increased correlation is favorable to the C/I distribution since whenever the received (wanted) signal is weak, the interfering signals will also be likely to be weak.

4.5 Directional Antennas and Sectorization

The systems that were analyzed in the previous section all used omnidirectional antennas. This is a quite natural solution since one strives to achieve a good coverage of the service area. Further, a RAP usually has no information about where the mobile terminal is located. The disadvantage of an omnidirectional antenna system is (besides providing low antenna gain) that the access port radiates interference power in all directions—not only the wanted direction. If directional transmitting antennas could be employed, the interference levels could be substantially reduced. In addition, a directional receiving antenna would be able to suppress interference from unwanted directions. This could, in turn, result in lower reuse distances being used (i.e., higher capacity). In this section, the effects of using directional antennas will be briefly analyzed.

Directional antennas tend to be rather bulky and are for this reason most commonly used at the RAPs. The antennas can be fixed, or adaptive. RAPs using fixed directional antennas, will usually require an array of antennas, each covering a sector with apex at the RAP. A mobile connected to the RAP moving in the service area is served by one of the directional antennas. As the mobile moves, a hand-off to a different antenna may be

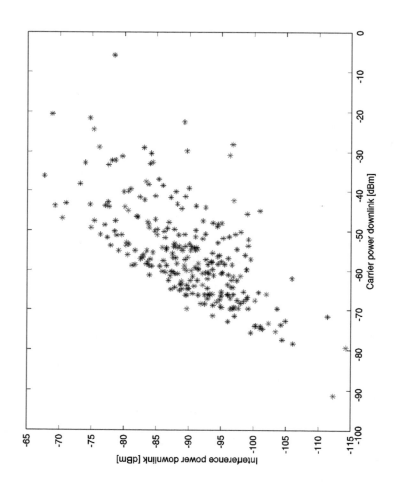

Figure 4.16 Correlated shadow fading. RUNE model simulation. $\alpha = 4$, $\sigma = 8$ dB. $K = 9$. Omni antennas. Left graph: Scatter plot of downlink wanted signal power vs interference power for correlation $\rho = 1$. Right graph: CDF of downlink C/I. Correlation (from left) $\rho = 0, 0.5, 1$.

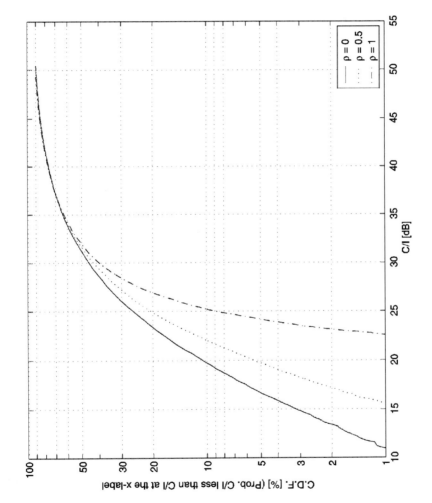

Figure 4.16 (continued).

necessary. An adaptive antenna array on the other hand would continuously track the mobile terminal. This has been shown to be a very effective tool both for mobile and fixed wireless systems. In particular, modern signal processing applied to electronically steerable array antennas has in recent years, proven useful in this context. The latter lies beyond the scope of this book. The analysis here will be confined to the simpler, fixed antenna case. Most of the benefits, will however, appear already in this case.

For this purpose, assume that the RAP are equipped with ideal *sector antennas* which in the horizontal plane of the far-field has the relative intensity

$$S(\varphi) \approx \begin{cases} \dfrac{2\partial}{\varphi_h} & |\varphi| \le \varphi_h/2 \\[2ex] \dfrac{1}{A_{sl}} & |\varphi| > \varphi_h/2 \end{cases} \tag{4.28}$$

The antenna radiation diagram [7] is characterized by the horizontal lobe width ϕ_h and the side lobe attenuation A_{sl}. Note that an omnidirectional antenna would be described by $\phi_h = 2\pi$. The simple capacity analysis from Example 4.2 will now be reiterated to study the impact of the directional antennas on the interference situation. Again, it is assumed that the RAP is located at the center of the cell and points (one of) its antenna in the direction $\phi = 0$ (see Figure 4.17). The uplink SIR can now be written as

$$\Gamma_j^u = \frac{P_j G_{i_0 j} S(\phi_0)}{\displaystyle\sum_{k \in M^{(c0)}} P_k G_{i_0 k} S(\phi_k)} \tag{4.29}$$

Assume that the terminal communicating with the RAP is within the main lobe of the antenna. It may be noted that signals from those interfering terminals (using the same channel) which are located outside the main lobe are (strongly) attenuated. However, signals from terminals within the main lobe (shaded area in Figure 4.17) are received with higher power than in the omnidirectional case. However, since the desired signal experiences the same antenna gain, the relative interference power of these latter terminals remains the same as in the omnidirectional case.

If the side lobe attenuation is large (4.24) can be rewritten as

$$\Gamma_j^u = \frac{G_{i_0 j}}{\displaystyle\sum_{k:|\varphi_k| < \varphi_h/2} G_{i_0 k}} \tag{4.30}$$

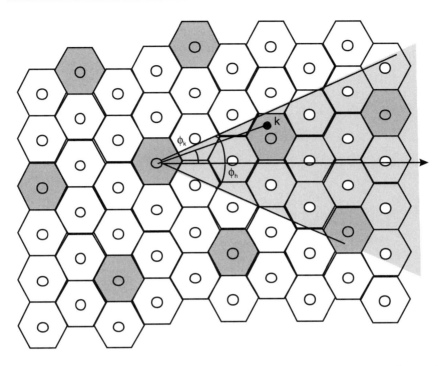

Figure 4.17 Geometry and interference analysis in systems with directional (sector-) antennas.

Compared to the omnidirectional case, the SIR increases in the same proportion as the effective number of interfering terminals (those falling into the main lobe). As a crude approximation, one can say that an antenna with main lobe width $2\pi/N$ reduces the average interference power by a factor N. In practice one is also interested using directional antennas to reduce the number of RAP sites and thus reduce the infrastructural costs of the system as illustrated in Figure 4.18. A typical cellular telephone system colocates 3 RAP (base stations) at the same site. The site is located at the corner of 3 cells, where each RAP uses a 120°-sector antenna to illuminate its cell. Otherwise, the cellular pattern remains the same. Thanks to the directional antennas the interference power is reduced since typically only 1/3 (= 120/360) of the interferers are visible to the RAP. It can be shown that for reuse factors divisible by three, it is possible to find frequency assignments where the nearest cochannel neighboring RAPs are not facing each other (thus avoiding the worst source of interference). Such reuse patterns are denoted $x/3x$ where x denotes the reuse factor for number of

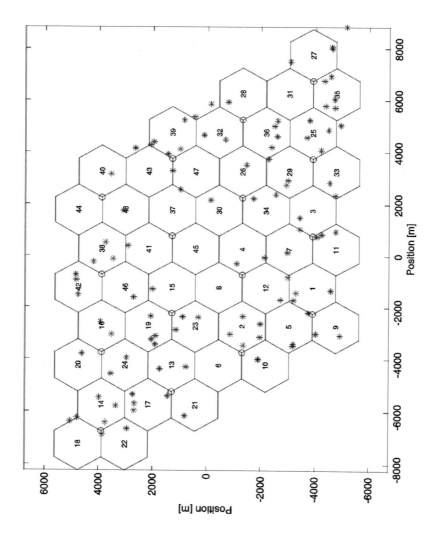

Figure 4.18 120 degree site pattern and outage probability for 1/3, 3, 9, and 3/9 reuse.

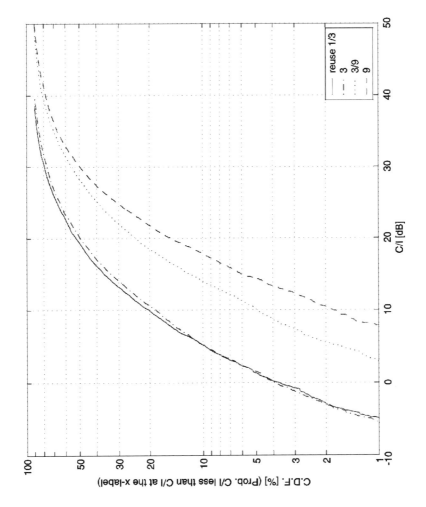

Figure 4.18 (continued).

RAPs (i.e., how many different RAP channel assignments there are, whereas the 3x denotes the reuse factor for cells. Common examples are 1/3, 3/9 and 4/12. In the 1/3 case, all RAP use all frequencies but split the channels into three sets, one for each of the sectors. On the other hand, placing a RAP at the corner of a cell will cause a terminal to be up to twice as far from its RAP as in the cell-center case. As can be seen from the CDF plot in Figure 4.18, there is a slight loss in performance for the sectorized system. The following example analyzes such a system using the simple analytical tools in the following example.

Example 4.7

Assume that the RAPs in the system in Example 4.2 use ideal 120°-sector antennas, located at the corner of the cells. Further, we assume that all base stations using a certain channel group illuminate their cells from the same direction. Estimate the η that can be used if the other requirements are the same as in Example 4.2 (no fading).

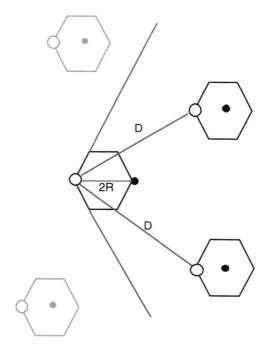

Figure 4.19 Geometry in Example 4.7.

Solution:

The worst interference case occurs as a terminal is at its largest distance from the RAP, that is, at distance $2R$. The 120° sector antennas will receive interference from only two of the six nearest co-channel neighbors (Figure 4.12). Since all RAP-antennas are turned the same way, both interfering terminals have to be located "behind" the serving RAP as seen from the studied cell. We approximate the distance from these mobiles to the RAP under investigation as $D + 1.5$. Considering only the first tier of interferers, the SIR can now be expressed as

$$\text{SIR} \approx \frac{\dfrac{cP}{2^4}}{2\dfrac{cP}{(D + 1.5)^4}} = \frac{(D + 1.5)^4}{2^5} > 100 \ (20 \ \text{dB})$$

$$D > (3200)^{1/4} - 1 \approx 6$$

From the expression (4.6) we get

$$K' = \frac{1}{3}D^2 \approx 12 \text{ i.e. } K = 12$$

The number of channels per cell is

$$\eta = \frac{M}{K} = \frac{100}{12} \approx 8 \text{ channel/cell}$$

which is about the same capacity as achieved with omnidirectional antennas in Example 4.2. The number of RAP sites is, however, only 1/3 of the number used in the omnidirectional case.

Problems

(* RUNE simulation solution required/recommended)

4.1 Let us consider a cellular telephone system with 100 channels. The system cell layout consists of a regular hexagonal grid with the base stations placed in the corners of each cell. The base stations use directional antennas. It is easy to see that for the cases where K is divisible by 3, the base stations will illuminate cells with the same channel group from the same direction. Assuming that the base stations use ideal sectorized antennas with 60° lobe width, that is, every cell is divided into two halves requiring two channel groups. Compare the capacity (measured as channels/area unit) of this system with

a) A system with 120° sector antennas and

b) A conventional system with omnidirectional antennas and base stations in the center of the cell. The SIR requirement (including all fade margins) is at least 18 dB.

4.2 A cellular telephone system uses a static channel allocation with $\eta = 20$ channels per cell. The arrival process of new calls is modeled as a Poisson process with arrival rate $\lambda = 180$ calls/h/km^2. Every call has an exponentially distributed duration with an average $1/\mu = 2$ minutes.

a) Determine the traffic intensity (Erlang/cell) if the cell radius is $R = 1$ km?

b) What is the blocking probability at this traffic load?

c) Determine the channel assignment failure rate at the traffic load if we can assume that (almost) all calls are handled by the system?

4.3 We would like to design a cellular telephone system along a (straight) freeway. In order to achieve reliable communication in the uplink, our SIR requirement is 20 dB. Assume that we have free-space propagation conditions. The base stations are spaced at regular intervals with one base station each kilometer. How many channels are required in the system if we are to handle at least 10 active terminals in each of the cells along the freeway?

Hint: Make sure the SIR requirement is met on the cell boundaries (1/2 km from each base station). Assume for the sake of simplicity that the interfering mobiles are close to their respective base stations. We also know that

$$\sum_{n=1}^{\infty} \frac{1}{n^2} = \frac{\pi^2}{6}$$

4.4 In a cellular telephone system, 90 channels are assigned to a straight road. Base stations are placed along the road with 2 km in between two stations, only radiating signals in one of the directions of the road. The channels are assigned according to a static cell plan. Assume that the propagation loss increases with the fourth power of the distance.

a) Determine the capacity (in channels/km) that can be achieved if the SIR requirement is 23 dB?
b) Determine the blocking probability as a function of the relative traffic load (Erlang/cell/channel).
c) Determine the channel assignment failure rate as function of the relative load (active terminals/cell/channel).

4.5 Consider a sector antenna for mobile communications with beamwith ϕ_H in the horizontal plane and beamwith ϕ_V in the vertical direction (i.e., covering (0, ϕ_V) in elevation). In the main beam the intensity can be assumed to be constant. Off the main directions there are many small side-lobes. The radiated energy in these directions, which are modeled with a constant intensity, are 20 dB lower that the main lobe intensity. Determine the antenna gain of this antenna. What is the gain of an omnidirectional antenna of this type ($\phi_H = 180°$)?

4.6 In a small wireless local loop (WLL)-system a square service area is served by 9 base stations placed as in Figure 4.20. The system uses single frequency TDMA and directional antennas where the antennas are adapted in each time slot. Assume that the base stations are using ideal sector antennas with beamwidth 5°. Assume that the terminals are uniformly distributed over the area and that the propagation loss is distance dependent only following a 4th-power law. The minimum required SIR is 25 dB and the noise can be neglected.

Determine the outage probability in the downlink for terminals at distance $R/2$ as function of the slot activity factor q. What happens if the SIR requirement is increased?

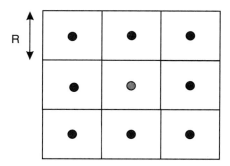

Figure 4.20 Problem 4.6.

4.7 In a wireless local loop (WLL)-system a square service area is served by base stations on an hexagonal grid. The system uses single frequency TDMA and directional antennas where the antennas are adapted in each time slot. Assume that the base stations are using ideal sector antennas. Assume that the terminals are uniformly distributed over the area and that the propagation loss is distance dependent following a 3rd-power law and is subject to log-normal shadow fading. The minimum required SIR is 25 dB and the noise can be neglected

a) Determine the outage probability in the downlink for a randomly chosen terminal as a function of the channel activity q for beamwidths 5°, 10°, and 20°. Assume that the closest base station is selected as the serving base station.

b) Assume that the terminals are using fixed directional receiving antennas of the same type as the base station. The antennas are assumed to be pointed directly at the closest base station. Repeat a) for the three beamwidths for this case.

Assume that now every base station is using two steerable antennas that can be simultaneously steered toward two different terminals in each slot, provided they are angle-wise not to close. Assume that the base station also otherwise is capabable of maintaining two connections in each slot. Repeat a) for this case. (Note that q in this case should be interpreted as the average number of terminals per slot, that is, from zero to two in this case.)

4.8* An area of 1000 km^2 is to be covered by a cellular radio network. Due to cost reasons, it is only possible to locate the base stations at 120 sites. 150 channels are available for the system. The lowest acceptable SIR is 14 dB and that requirement must be met over the whole area. The blocking probability should not exceed 2%. The path gain can be considered to be proportional to $1/r^4$. Estimate the area capacity (Erlang/km^2) gain/loss when using directional antennas with 120 degree lobes instead of using omnidirectional antennas?

Hint: Effects at the border of the system can be disregarded.

4.9* In a wireless local loop (WLL)-system a square service area is served by base stations on an hexagonal grid. The system uses single frequency TDMA and directional antennas where the antennas are adapted in each time slot. Assume that the base stations are using ideal sector antennas. Assume that the terminals are uniformly distributed over the area and that the propagation loss is distance dependent following a 3rd-power law and is subject to log-normal shadow fading. The minimum required SIR is 25 dB and the noise can be neglected.

a) Determine the outage probability in the downlink for a randomly chosen terminal as function of the channel activity q for beamwidths 5°, 10°, and 20°. Assume that the closest base station is selected as the serving base station.

b) Assume that the terminals are using fixed directional receiving antennas of the same type as the base station. The antennas are assumed to be pointed directly at the closest base station. Repeat a) for the three beamwidths for this case.

c) Assume that now every base station is using two steerable antennas that can be simultaneously steered toward two different terminals in each slot, provided they are angle-wise not too close. Assume that the base station also otherwise is capabable of maintaining two connections in each slot. Repeat a) for this case. (Note that q in this case should be interpreted as the average number of terminals per slot, that is, from zero to two in this case.)

4.10* In a densely packed analog cellular mobile telephone system base stations are placed on a regular hexagonal grid. The distance between base stations is 2 km. The total available number of channels is 400. A static assignment strategy is used with a symmetric hexagonal cell plan with base stations using omni-directional antennas at the center of each cell. The modulation scheme requires minimum SIR of 15 dB and slow handoff is used. Assuming that the traffic is uniformly distributed, what is the capacity of the system (in Erlang/km^2)

a) If a blocking probability of 5% and an outage probability of 5% can be accepted, and

b) If the GOS, defined as the sum of outage & blocking probabilities are not to exceed 10%? The operator using a design according to b) is now forced to surrender half of its bandwidth to competing digital services, that is, only 200 channels remain.

c) What is the capacity in the system now?

4.11 A 1-D cellular system has 80 channels at its disposal. The modulation scheme employed requires a minimum SIR of 14 dB. Determine the minimum distance between radio access ports if the channel assignment failure rate is not to exceed 1% and the wanted area capacity is 5 mobiles/km. Assume that the pathloss increases as the alpha = 2.5 power of the distance.

Hint: Make and motivate necessary approximations.

4.12   If cellular radio systems covering a desert highway with the RAPs are located along the straight road, each RAP has 40 channels at its disposal. A modulation and coding scheme requiring 19 dB SIR is employed. The required area capacity is 10 mobiles/km for a channel assignment failure rate of 2%. How many fewer RAPs are needed if we use directional antennas (radiating only in one direction along the road) instead of omnidirectional antennas (radiating in both directions)? Assume that the propagation loss increases with the 4th power of the distance.

4.13   In a cell, terminals are uniformly distributed over the surface. Derive the probability density function of the distance r of a terminal from the center of cell if
a)   The cell is circular; and
b)   If the cell is hexagonal.

References

[1]   Lee, W. C. Y., *Mobile Communication Fundamentals*, New York: Wiley, 1993.

[2]   Leese, R. A., "A Unified Approach to the Assignment of Radio Channels on a Regular Hexagonal Grid," *IEEE Trans. Veh. Tech.*, Vol. 46, No. 4, Nov. 1997.

[3]   Cox, D. C., "Cochannel Interference Considerations in Frequency Reuse Small Coverage Area Radio Systems," *IEEE Trans. Commun.*, Vol. 30, No. 1, 1982.

[4]   MacDonald, V. H., "The Cellular Concept," *Bell Syst. Tech. J.*, Vol. 58, No. 1, 1979.

[5]   Schwartz, S. C., and Y. S. Yeh, "On the Distribution Functions and Moments of Power Sums with Log-Normal Components," *Bell Syst. Tech. J.*, Vol. 61, No. 7, 1982.

[6]   Beaulieu, A. A., et al., "Estimating the Distribution of a Sum of Independent Log-Normal Random Variables," *IEEE Trans. Commun.*, Vol. 43, No. 12, 1995.

5

Handover and Mobility

5.1 Mobility Management Fundamentals

Mobility provides the possibility to communicate in different locations and while on-the-move, and is one of the key characteristics of wireless networks. This is made possible by the fact that terminals are not attached to the fixed infrastructure by wires. Mobility over a few meters is clearly not a big problem. More interestingly, what happens if the terminals move larger distances, for example, when users drive around in their cars or travel long distances by aircraft and switch on terminals far from home, possibly on a different continent? Two types of problems arise in this context: tracking inactive terminals in order to be able to respond quickly to requests from the fixed network to establish communication with the terminal. The other problem is that a moving, active terminal will face the risk that it will leave the area where its current radio access point is capable of providing sufficient QoS. The former problem is of global scale and does not only affect a single network operator, since a terminal may well leave the service area of the operator and enter the service area of another operator. This is usually referred to as the *roaming* or *locating* network functionality. The latter problem is of real-time character and is particularly demanding when it comes to circuit switched traffic, where seamless service is required with little or no loss of transmitted data is permissible.

The roaming problem is mainly a fixed network issue and is currently solved by maintaining central databases containing the current locations of all terminals, the location registers (LR). Whenever a session is being set up, current location data (and other user characteristics) are retrieved from the

LR and used to find the appropriate radio access point to serve as the wireless connection point. There is one important aspect, however, that affects the radio resource allocation, which we will briefly discuss here. Since terminals may be moving at all times, the accuracy of the location data may vary and will be dependent on the rate at which this information is updated. This is of course most problematic for the majority of terminals that are not currently active and thus are not transmitting any data, nor are they attached to any access point in particular. In order to update the information, these terminals will be required to transmit location update messages to the network in order to report their location. In fact, the physical location may be neither easy to obtain, nor may even be required.

In most systems, inactive mobiles will be required to monitor the signal-levels received from nearby access ports and report which of them are likely candidates to become servicing access ports if a session involving the terminal will be requested by some other party. If the reporting occurs at slow rate, there is a fair chance that the terminal has moved considerably when the LR-data is being requested. In order to establish the session, search or paging messages have to be transmitted not only from the previously reported candidate access point, but also by several neighboring access points. The size of the area where the search has to be performed is directly related to the rate of location updates and the mobility of the mobiles. From a resource management point of view, there is an interesting trade-off to be made: how much radio resources should be spent on transmitting location update messages from nonactive terminals (which can occur in large numbers) in order to conserve radio resources for paging terminals at session set-up.

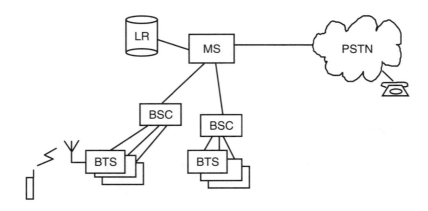

Figure 5.1 GSM architecture.

As an example, take the GSM system, where the service area is divided into location areas (LAs) containing several access ports (base stations). Each base station continuously transmits its identity and the location area it belongs to on its Broadcast Control Channel (BCCH). Nonactive terminals continuously monitor the identities of the surrounding base-stations and when the primary candidate base station for a would-be connection belongs to a different LA, a location update message is transmitted to this base station.

In addition, terminals transmit location messages when they are switched on and off (so called "attach" and "detach" messages). The operator will need to choose the size of the LAs. Large LAs with many base stations will result in few location update messages, but will require a large number paging messages (over all the base stations in the LA) in order to find the terminal. Small LAs will, in turn, give rise to frequent location updates, but result in precise and fast session setups with little waste of radio resources in the signaling channels.

Providing seamless service to active users while data transfer is in progress—the instantaneous selection of the serving access port—is, on the other hand, a problem that directly relates to radio resource management (see Chapter 3). The handover or handoff can be divided into three phases or subproblems:

1. *Handover decision/detection*—when and where to handover: In the first phase, a decision is made to initiate a handoff. This decision is usually based on measurements of the current link quality or predictions of the future link quality. Measurements can be made by the terminal or the access ports. The terminal or the network

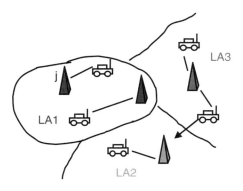

Figure 5.2 Roaming and location areas.

can make the decision to initiate a handover and decide what will be the destination access port of the handover procedure. Most cellular telephony systems (e.g., GSM) rely on so called Mobile-Assisted (Network Controlled) (MAHO) handover, where the mobile continuously measures and reports signal levels and received signal quality to the network (BSC) where the decision to initiate the handover is made. Mobile controlled handover is another mode of operation used in practice (e.g., DECT). This problem is addressed in more detail in Section 5.2.

2. *Handover resource assignment*—are there resources available in receiving access ports? When a (quality-based) decision to handoff has been made and a target access port has been determined, the question arises whether there are enough radio resources in the receiving access port. The most obvious question in an orthogonal wave-form system is, of course, if there is a channel available to receive the terminal. An example of a more intricate constraint is protecting the link quality in already ongoing connections in other access ports in the vicinity using the same channel as selected for the inbound handoff connection. In such cases, the handoff may be denied (even though there is a channel available) and either the current access port has to be kept, or a new candidate access port for hand-over has to be selected. This problem is studied in more detail in Section 5.3.

3. *Hand-over execution*—protocols for the reliable exchange of hand-over data: In this last step a target access port has been selected that is capable of receiving the terminal. Now a signaling procedure to facilitate the handover is needed to inform the involved terminal and access port about the new resource allocation. This may also require synchronization procedures. The exchange has to be swift and reliable in order not to lose payload data while the actual switch-over is done, and not to drop the connection.

5.2 Handover Decision Algorithms

During a session, a mobile terminal can communicate with the same access port, as long as the signal quality is good enough. In traditional mobile telephony systems the coverage area of an access port is rather large and during a phone call not many users actually experience a handover. As the demand for capacity increases and more and more small cell (microcellular) systems are introduced, the requirement on handoff reliability increases as many more handoffs are expected. This is because a mobile may pass more cell boundaries during a call or session. The decision to initiate handoff is a vital component in the process since the success and the efficiency of the handoff mechanism, to a large extent, depends on the accuracy and timeliness of the decision. Most cellular telephony systems (e.g., GSM) rely on MAHO handover, where the mobile continuously measures and reports signal levels and received signal quality to the network (BSC), where the decision to initiate the handover is made.

The decision is usually based on measurements of the current link quality or predictions of the future link quality. Measurements can be made either by the terminal or the access ports. In this section it is assumed that such measurements are available in the decision process. Normally, a mobile terminal will have several access points that could be candidates to become the receiving access point in a handoff. For the sake of simplicity, the situation when there are only two access points involved will be considered, one serving access point and one (potentially) receiving access point. This situation is depicted in Figure 5.3. The figure also shows typical signal quality measurements from the two access ports A and B as function of the mobile location. For the time being, the fact that the signal quality also will be time-dependent due to the fluctuations of the cochannel interference will be neglected. In a classical macro-cell system with a nonaggressive cell plan (high reuse factors K), the approximation will be quite good, since the signal quality will be

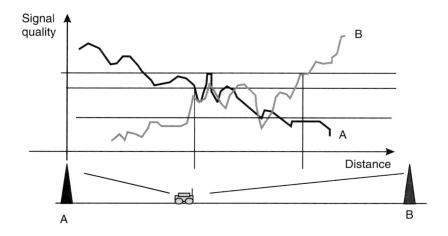

Figure 5.3 Signal quality based handover detection.

dominated by the wanted signal level (which is only location dependent). Making this assumption, we can model the signal quality by the received signal strength (RSS). Assume further that the RSS from access point i is sampled at time instants t_k as $S_i(t_k)$. A handoff algorithm can be described by an integer-valued function $B(t)$ that, given all the observed RSS samples, indicates the access port that should be selected,

$$B(t_k) = f(S_1(t_k), S_1(t_{k-1}) \ldots , S_2(t_k), S_2(t_{k-1}) \ldots ,$$
$$S_B(t_k), S_B(t_{k-1}) \ldots)$$

The analysis here will be confined to a certain category of handoff decision algorithms, defined as

Definition 5.1:

A handoff decision algorithm is said to be *separable* if the decision function B can be written as:

$$Y_i(t_k) = f_i(S_i(t_k), S_i(t_{k-1}) \ldots)$$
$$B(t_k) = f(Y_1(t_k), Y_2(t_k) \ldots Y_B(t_k))$$

Definition 5.2:

A handoff decision algorithm is said to be *separable* and (*log-*) *linear* if the decision functions Y_i can be written as

$$Y_i(t_k) = H*A^T$$
$$A_{ik} = 10 \lg(S_i(t_k))$$

where H is a square weighting matrix with the same dimension as the number of observations. The decision variables Y_i are thus derived as linear combinations of the measured (estimated) RSS decibel-values.

There are several ways to assess the performance of a handoff decision scheme. Two of them are considered below.

1. *Dropping probability* is a measure of how often a handoff fails. A failure occurs if the required signal quality is below the required value for more than a given time interval. The length of this interval is highly dependent on the application/service provided by the

wireless network. A real-time, delay sensitive service, such as voice, can tolerate only short interruptions and after a few seconds the call may be considered to be lost. In nonreal time data services a longer interruption may go completely unnoticed since no, or only a few, packets were transmitted during the time when the quality was too low. Also, the procedure to complete the handoff during the execution phase comes into play. A long and sensitive procedure may be prone to errors or delays such that the connection may be lost before the signaling procedure to hand off the mobile fails. This is due to poorer and poorer signaling quality during the handoff execution phase. In this study a simplified measure is used and only outage probability during the handoff period is considered, that is, the probability that the signal quality is too low during this period.

2. *Handoff probability/rate* is a measure indicating how often handoff decisions are made. Looking at Figure 5.3, if the mobile moves from access port A to access port B, in principle only one handoff has to be made. Due to the strong fluctuations in signal quality it is obvious that several more handoffs may be executed, some of them necessary to maintain a high enough signal quality, some not necessary at all. The excessive handoffs will put an extra burden on the signaling system of the network as well as constitute a risk of loosing the connection, if the handoff execution procedure fails. The number of handoffs is also dependent on the rate at which measurements are available and how often handoff decisions are made.

We can guess that there is some kind of tradeoff between the handoff rate and the dropping probability. Make too few decisions and handoffs increase the risk of dropping the call. On the other hand making many decisions will increase the handoff rate (and, in addition, load the signaling channels with measurement data). In order to assess this tradeoff, a better model of the signal level fluctuations is needed. Assume as shown in Appendix A, that the received signal level can be written as:

$$S_i(r_i) = c \frac{G_i(r_i)}{r_i^{\alpha}}$$

where $G_i(r)$ are a log-normal random processes. In dB-scale we have

$$A_i(r) = 10 \lg(S_i(r)) = X_i(r) + C - 10\alpha \lg(r_i)$$

where $X_i(r)$ are zero-mean Gaussian random processes and C is some constant. The log-normal shadow fading will certainly be correlated such that $G(r)$ takes similar values for nearby distance values r. A simple model for this correlation was first presented in [2] and is outlined in Appendix A.

Now return to the performance tradeoff in the following example where three simple (hypothetical) handoff decision schemes are investigated:

Example 5.1

Assume that the mobile terminal in Figure 5.3 is moving at constant velocity of 20 m/s on the straight line from access port A to access port B. The access ports are separated by 1,000m. The signal level decays with the 4th power of the distance and at the midpoint between the access ports the average signal level from each of the access ports is $\Delta = 5$, and 10 dB above the minimum required level to maintain the connection. The log-normal shadow fading has a standard deviation of 8 dB and the correlation distance (where the correlation has dropped to 0.5) is 50 m. Signal level samples are available every 0.5 seconds and handoff decisions can be made at any sample point. Handoff execution is assumed to be immediate and error free. Determine

a) The outage probability (i.e., the probability that the received signal level is below the threshold at some point in the handoff process)

b) The expected number of handoffs for the three cases:

I.   Instantaneous handoff: The access port with the highest signal level is chosen in any sample point;

II.   Averaging: The maximum average signal level (over the last 10 samples) determine the selected access port;

III. Ideal averaging: The true expected value of the signal level is assumed to be known and the access port with the highest expected signal level is selected.

Solution:

The signal levels received from the two access ports (in dB) are:

$$A_1(r) = X_1(r) + C - 40\lg(r)$$
$$A_2(r) = X_2(r) + C - 40\lg(1000 - r)$$

where X_i are independent Gaussian random processes. Using the midpoint criteria, that is, $E[A_i(500)] = A_{min} + \Delta$ (dB) yields:

$$C = A_{\min} + 40 \lg(500) + \Delta \ (\text{dB})$$

Figure 5.4 shows a typical realization of the three algorithms. As we can expect, algorithm I is equivalent to selection diversity and does the best. The averaging algorithm does reasonably well, but exhibits the typical lag behavior as it keeps selecting access point A even after passing the 500m mark (sample 50) with an outage situation in the region sample 50–65. We have that

$$a = \epsilon_D^{vT/D} = 0.5^{20-0.5/50} = 0.5^{0.2} \approx 0.87$$

A simple MATLAB-simulation is used to estimate the probabilities and the results are shown in Table 5.1. Note the importance of the handoff margin Δ. A large margin lowers the outage probability quite drastically. As one may expect, the strongest instantaneous signal provides the best results concerning the outage probability, whereas the ideal handoff performs the worst in this respect.

It is also clear that the expected number off handoffs, R_{OH} is very large for the instantaneous scheme. For the ideal handoff scheme, III, exactly one handoff will occur (always at the midpoint). In [1], some analytical tools are available to calculate some of these parameters using the fact that the shadow fading signals measured in dB are Gaussian processes.

The example above shows clearly the importance of signal processing to strike the correct balance between the number of handoffs and the outage probability. Here a simple signal-level averaging scheme is used, but obviously better techniques utilizing knowledge about the fading environment could be employed. A common strategy employed in the literature is to design a prediction filter to predict those situations in advance where the signal quality would deteriorate beyond salvage and make a (necessary) handoff prior to that. The drawback with good prediction techniques is that the general model must fit and the parameters of the fading have to be known— something that is hardly true in practice. Also, simple nonlinear techniques from control theory such as signal hysteresis are commonly used to reduce the flip-flopping back and forth between access ports and thus an excessive number of handoffs. In these systems the handoff decision threshold changes once a decision is made such that the required signal level difference for changing back is now higher.

In the previous discussion, the received signal level was used as the handoff criterion. This is, as can be seen in the next example, not always unproblematic, since the interference may vary considerably in practice.

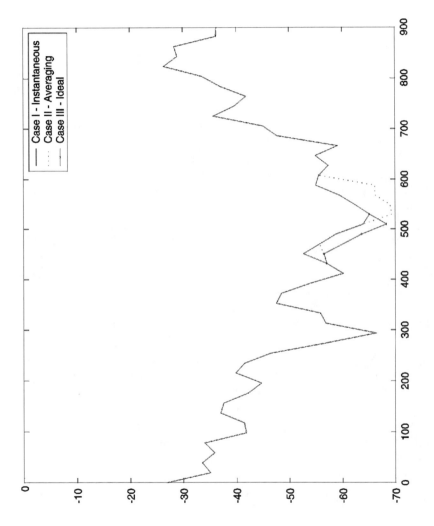

Figure 5.4 Typical realization of resulting signal levels for the three algorithms (I blue, II green, III red).

Table 5.1
Solution of Example 5.1

Algorithm	$P_{out}\Delta = 10$ dB	$P_{out}\Delta = 5$ dB	$P_{out}\Delta = 0$ dB	R_{HO}
I	0.3%	2.4%	9%	7.6
II	1.4%	5%	13%	1.8
III	2.0%	6%	14%	1

Example 5.2

Let a mobile terminal move in a cellular system as shown in Figure 5.5a. The signal level decays with the 4th power of the distance and the shadow fading has a standard deviation of 8 dB and a correlation distance of 50 meters. Plot the signal level, the interference level, at some randomly chosen channel at the receiving access port and the resulting signal to interference ratio.

Results:

Figures 5.5b–c show the required results. A SIR-based scheme would make one handoff at around 800 meter whereas the signal level-based schemes would probably make several handoff back and forth. Figure 5.5b–d shows that the signal-to-interference ratio does not necessarily follow the signal level. In this particular case the channel(s) of the initial access port is (are) subject to a much higher interference level in the interval 1,000–2,000m which would make an early handoff (around 500m) much more favorable than one could realize from the signal level graph alone.

The difficulties of using the signal level alone as detection criterion is further highlighted when studying more extreme propagation conditions such as in the Manhattan model outlined in Appendix A. This was first discussed in [3], where the handoff in the presence of rapid signal level changes due to mobile terminals turning around street-corners was investigated. This type of problem is illustrated by Figure 5.6. Here it is assumed that a mobile terminal moves on a street close to access point BS1 in the lower middle of the left map upwards in the graph. The signal level and the SIR are plotted for the two cases where the mobile goes straight towards the far away access port BS3 (case A) or takes a right turn towards access port BS2 (case B). Assume further that the signal levels are reasonably high and that BS3 is reusing the same channels as BS1, whereas the other two access ports are

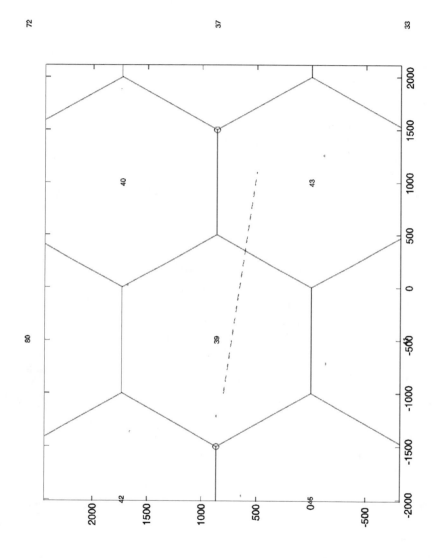

Figure 5.5 Example 5.2. Mobile trajectory.

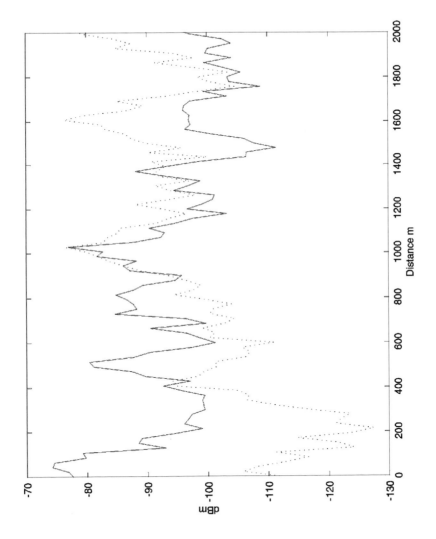

Figure 5.5 (continued). Received signal-level.

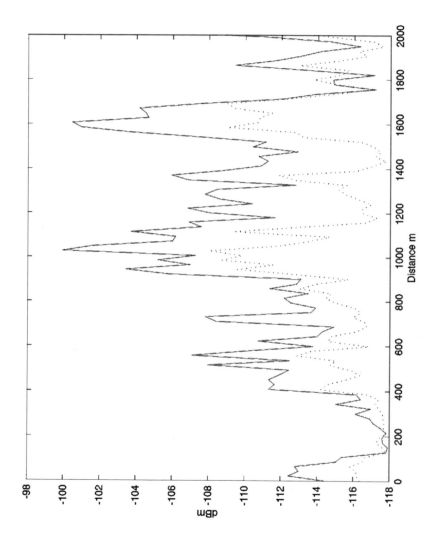

Figure 5.5 (continued). Interference levels.

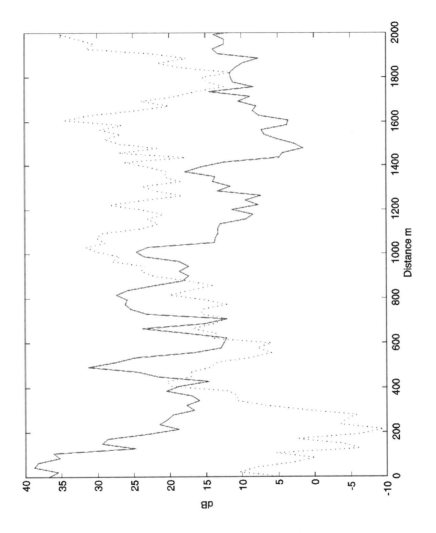

Figure 5.5 (continued). Signal-to-interference levels.

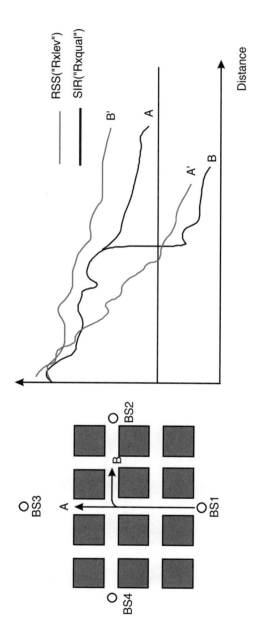

Figure 5.6 Received signal levels (RXLEV) and signal-to-interference ratio (SIR) in Manhattan-style propagation environments.

not. Now, for the A-case, note that although the signal level (A) drops moderately, the SIR (A') drops very quickly. A signal level based scheme would in this case underestimate the urgency of a handoff decision. For the case B, where the mobile takes the turn, the graph shows a dramatic drop in signal level (B) as the mobile moves into the building shadow immediately after the turn. A signal level based decision would be to make a handoff as quickly as possible. Looking at the SIR (B') gives a completely different picture, though. Since the main interferer is BS3, moving around the corner will cause both the signal level and the interference level to drop about the same amount, leaving the SIR almost unchanged. A handoff is certainly not needed from an SIR point of view.

The example above shows that the received signal level may not always be trusted as a handoff criterion, in particular, not in high-density systems where the coverage is not a major problem, but interference is. In addition, in these systems it is difficult to estimate the signal level of the wanted signal, since only the total (signal plus interference) level can be measured directly. A high total received level could also mean that the interference level is high, not only that the wanted signal is strong. In practical systems a combined decision criteria is used.

In the GSM system [6], the handoff decisions are based both on signal level measurements on the current link, and potential receiving neighboring base stations and signal quality estimates on the current link. Since the SIR is difficult to measure directly, the raw bit-error rate of the received information is used to produce the RXQUAL measure, which is quantized to 8 levels (3 bits). When a call is set up, mobile terminals are required to report averaged signal levels (RXLEV quantized to 64 levels corresponding to 6 bits) and RXQUAL values every 0.5 sec on their active link. This information is multiplexed with the voice traffic data on the so-called slow associated control channel (SACCH). The exact algorithm for handoff decisions is not specified in the GSM specification and each system's manufacturers may use these measurements in their own way to make the handoff decisions. The specifications (GSM 05.08, Annex A) provide an illustrative example on how a typical handoff scheme could be designed. In principle, the 32 last samples in the interval between two reporting instants of the RXLEV and the RXQUAL measurements are stored and averaged for both the terminal and the base stations. These averages are then compared with threshold values. If a certain number of successive samples of each of the average measurements are below the threshold, a handoff-required situation is reached. Also, criteria, such as distance from the base station (measured by the delay) and the transmitter power level can be used to reach this state.

During normal operation of a link, the signal levels from other candidate base stations are also collected and stored. A base station candidate list is compiled and for each base station entry in this list a criterion function is evaluated. A straightforward criterion function may be the path loss, that is, the used transmitter power minus the received signal level (in dB). The candidate list is then sorted in decreasing order according to the criteria function values (e.g., the path loss) setting the candidate with the best value (e.g., the lowest path loss) first. Normally the top candidate would be selected to become the receiving base station in the handoff procedure. In addition to these criteria, other parameters such as priorities and traffic load conditions can be used to select other candidates rather than the top one.

5.3 Handover Resource Management

Once a handoff decision has been made, a receiving access port has to be determined. As in the GSM example, the candidate to become the receiving access port can be selected on the grounds of signal quality. Now the question arises whether there are enough radio resources available in the receiving access port. This could involve the availability of channels. Or the link quality in cochannel links in the vicinity of the receiving access port may become too poor to allow the handoff. In such cases the handoff may be denied and either the current access port has to be kept, or a new candidate access port for handover has to be selected. It is clear that terminals arriving to an access port as a consequence of a handoff compete for the same radio resources with terminals having selected that particular access port in trying to set up a new session. In many applications protecting the ongoing sessions is considered more important than denying new terminals access to the system. In a classical mobile telephony system, it is found to be more detrimental to the users to lose an ongoing call than to block a newly arrived call. The latter reasoning makes it obvious to reserve resources for handoff mobiles such that these resources (e.g., channels, time-slots) can only be used by terminals performing a handoff and not by newly arriving call set-up requests in the own cell. Such an arrangement is evaluated in the following example.

Example 5.3 Channel Reservation and Cost of Mobility

In a cellular telephone system, each cell reserves n channels for handover traffic in such a way that no new calls are admitted (i.e., only handover calls are allowed) when the number of channels in use is larger or equal to $\eta_0 = \eta - n$ channels. Otherwise the new call is blocked. Incoming handover calls are dropped only if no channel is available. New calls and handover calls can be assumed to arrive as independent Poisson processes with intensities λ_N and λ_H respectively. Calls have a lifetime in the cell, that is, they are terminated or leave the cell within a time interval that is exponentially distributed with a average $1/\mu$. Determine the blocking and dropping probabilities of a function of the traffic load when 50% of the total calls arriving at the cell are handovers and a total of $\eta = 12$ channels are available!

Solution:

Denote the total number of calls in progress in the cell at time t, $N(t)$. Note that whenever a call has arrived and is assigned a channel, it is no longer of any consequence to the number of calls in the cell, whether this call originally was handed over to the cell, or if it was a new call arriving at that cell. Due to the memory-less properties of the Poisson arrivals and the exponential distribution of the call life time in the cell, $N(t)$ will be a Birth-Death Markov chain with the state-transition diagram in Figure 5.8.

We proceed to derive the stationary state probabilities [14].

$$p_k = \Pr[N(t) = k]$$

by means of the flow-cut equations:

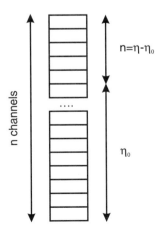

Figure 5.7 Channel reservation in Example 5.3.

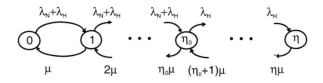

Figure 5.8 State diagram in Example 5.3.

$$(\lambda_N + \lambda_H)p_{k-1} = k\mu p_k \quad 1 \le k \le \eta_0$$

$$\lambda_H p_{k-1} = k\mu p_k \quad \eta_0 < k \le \eta$$

Iteratively solving these equations yields

$$
p_k =
\begin{cases}
p_0 \dfrac{(\lambda_N + \lambda_H)^k}{\mu^k k!} & k \le \eta_0 \\[3mm]
p_0 \dfrac{(\lambda_H + \lambda_N)^{\eta_0}(\lambda_H)^{k-\eta_0}}{\mu^k k!} & \eta_0 < k \le \eta
\end{cases}
$$

Using the fact that all p_k add up to unity, we can solve for p_0. Using the notation

$$\rho_{TOT} = \frac{\lambda_N + \lambda_H}{\mu}$$

$$\rho_H = \frac{\lambda_H}{\mu}$$

we get

$$
p_k =
\begin{cases}
\dfrac{\dfrac{(\rho_{TOT})^k}{k!}}{\displaystyle\sum_{j=0}^{\eta_0}\dfrac{(\rho_{TOT})^j}{j!} + \sum_{j=\eta_0+1}^{\eta}\dfrac{(\rho_{TOT})^{\eta_0}(\rho_H)^{j-\eta_0}}{j!}} & k \le \eta_0 \\[8mm]
\dfrac{\dfrac{(\rho_{TOT})^{\eta_0}(\rho_H)^{k-\eta_0}}{k!}}{\displaystyle\sum_{j=0}^{\eta_0}\dfrac{(\rho_{TOT})^j}{j!} + \sum_{j=\eta_0+1}^{\eta}\dfrac{(\rho_{TOT})^{\eta_0}(\rho_H)^{j-\eta_0}}{j!}} & \eta_0 < k \le \eta
\end{cases}
$$

Now, deriving the blocking and handover dropping, probabilities yields

$$P_{\text{block}} = \sum_{k=\eta_0}^{\eta} p_k$$

$$P_{\text{drop}} = p_\eta$$

Defining the relative mobility a as

$$a = \frac{\lambda_H}{\lambda_N + \lambda_H}$$

and evaluating the $\eta = 12$, and $a = 0.50$, Figure 5.9 provides the required numerical results. Figure 5.10 shows the dependence of the results on the relative mobility, whereas Figure 5.11 gives the grade of service (GOS) as

$$\text{GOS} = P_b + 10 P_d$$

for some different values of a. Studying the graphs, it is clear that there is a trade-off between blocking of newly arrived calls and handoff calls. If there is a requirement to keep the dropping probability low, more channels for handoff calls have to be reserved. As the relative mobility of the terminals is increased (more handoffs per call), the required number of reserved channels is getting larger and a smaller fraction of the channels can be used for newly arriving calls. As a consequence the capacity drops. At some given QoS level, a system with highly mobile users, or very small cells, has a lower capacity than a system where the mobility is lower. This is usually referred to as "the cost of mobility" [5].

The reservation policy has its limitations, in particular when the traffic load is high and mobile terminals are moving rapidly. On the other hand, in a high-density, high-capacity network, usually the list of candidates for receiving a session in a handoff may be quite long. If the (from a signal quality point of view) best candidate access port is chosen, the session may be rejected due to resource limitation. On the other hand, there may be several candidates that have acceptable (if not the best) signal quality and available resources (channels). Obviously, there is something to gain by making information about resource availability part of the handoff decision algorithm. One simple, but popular technique (used by several GSM operators), is traffic controlled handoff (TCH). In its simplest form, a signal level

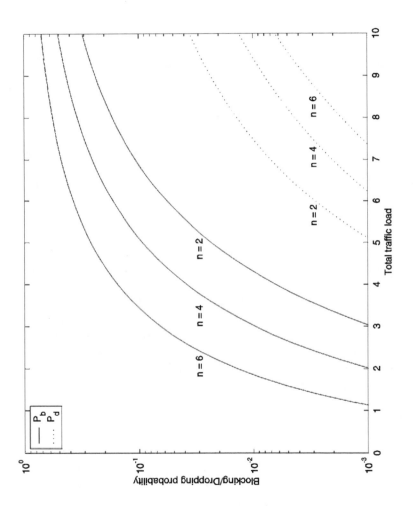

Figure 5.9 Blocking and handover dropping probability as function of total traffic load when *n* out of a total of $\eta = 12$ channels are reserved for handover traffic ($a = 0.5$).

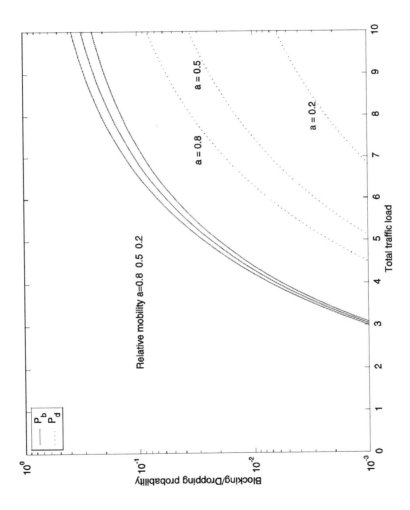

Figure 5.10 Blocking and handover dropping probability as a function of total traffic load when 2 out of a total of $\eta = 12$ channels are reserved for handover traffic for different values of the relative mobility a.

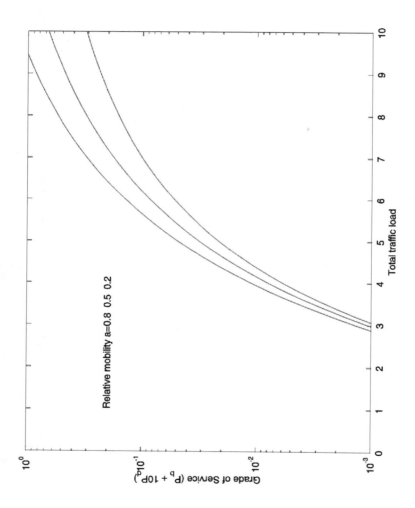

Figure 5.11 GOS $(P_b + 10P_d)$ as a function of total traffic load when 2 out of a total of $\eta = 12$ channels are reserved for handover traffic for different relative mobilites.

dependent handoff decision is made, but either the decision function or the decision threshold is allowed to be dependent on the current traffic load situation. An access port with a high traffic load will use a high signal level threshold (i.e., only those few mobiles that are very close are allowed to handoff to this access port) and conversely, a light load access port will decrease its threshold thus receiving a larger number of more distant terminals in handoffs. A very clever way to implement this is to vary the power of the access port control channel, which terminals use to assess the signal quality of the candidate access port. Heavily loaded access ports will lower their power, thus appearing to provide poorer signal quality and thereby becoming less likely candidates for handoff. Lightly loaded access ports will increase their power. Figure 5.12 provides a rough illustration.

As can be seen in Figure 5.12, the regions where acceptable signal quality is acquired from the respective access ports varies with the traffic load. The cells effectively shrink and grow over time. This phenomenon is also called "cell breathing" and is a popular technique in CDMA systems, which are covered in more detail in Chapter 9.

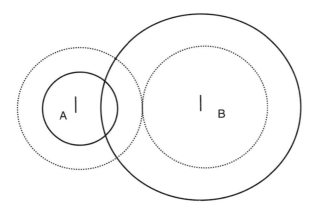

Figure 5.12 Traffic controlled handoff, "cell breathing" (dotted lines: normal region for accepting handoff sessions; solid lines: traffic dependent regions for accepting handoff—access port A is heavily loaded; access port B is lightly loaded).

Problems

5.1 In a system with hexagonal cells an omni-antennas mobiles perform handover according to a simple signal-level based strategy. The path loss is distance dependent (4th-power of distance) and exhibits log-normal shadow fading (σ = 8 dB).
   i) Fast (instantaneous) handoff: The mobiles are always connected to the strongest base station;
   ii) Slow handoff: The mobiles are connected to the base station with the highest (long-term) average signal power (i.e. closest base station).
On the boundary of a cell, estimate the ratio between the median field strengths in these cases. Neglect any noise and consider only the two closest base stations.

5.2 A terminal moves with a speed of v = 70 km/h through a wireless network with hexagonal cells. Assume that a handover is made each time per call if the average call duration is 1 minute and if the cell radius is 1 km and 50m respectively

5.3 In a cell in microcellular system with hexagonal cells, new calls arrive at a rate of 5 calls/minute (Poisson). There are two categories of terminals: vehicle-mounted moving at average speed 60 km/h and pedestrian, moving at an average speed of 5 km/h. The cell radius is 100m. Estimate the number of handoffs per minute if
   a) 80%
   b) 20%
of the calls are from pedestrian terminals.

5.4 In a cell in a microcellular system new calls arrive at a rate of 8 calls/minute (Poisson). In addition calls arrive to the cell (and others leave the cell) by means of handoff at a rate of 4 calls per minute. The average time a subscribe stays in the cell (i.e., until the call is either terminated or leaves the cell) is exponentially distributed with average 30 seconds. The total number of channels available is in the cell is 12.
   a) Determine the probability that a call is blocked or dropped.
   b) If 4 channels are always set aside only to be used for accepting handoff.
      Calls (i.e., no new calls are accepted when only 4 channels remain) what is now the blocking and dropping probability?

5.5 A small mobile telephone system covering a straight highway, there are four base stations operating at equal and constant power, located 1 km apart according to Figure 13. The area to be covered is the interval $0 \leq r \leq 4$ (km). We assume that each mobile is connected to the closest base-station. Assume that the number of mobiles is Poisson distributed with λ active terminals/length unit. Assume that the total number of channels in the system is large in relation to the traffic load. Mobiles are moving with constant but random velocity. The speed of the mobiles has a Gaussian distribution with mean and standard deviation (50, 30) km/h, with equal probability of moving right or left.
a) What is the expected time a mobile remains in a cell?
b) Plot the expected number of handovers from and to cell 2 as a function of the traffic load λ.

5.6 Microsurf plans to introduce a pico-cellular wireless indoor system. They measured at various customer sites that the access requests can be quite accurately modeled by a Poisson process and are estimated for the given cell size at 1 request per minute. Since most users have a notebook, hand-over from neighboring cells occur at the same rate. The duration is exponentially distributed with a 30 sec mean. The system has to comply to several European norms, which define the following requirements:
- Single-cell (i.e. no neighbouring cells) blocking probability < 0.1
- Multi-cell (i.e. cells have neighbours) blocking probability < 0.3
- Multi-cell dropping probability for handover < 0.1

The Microsurf system is designed to use 2 channels per base station (cell).
a) Is this sufficient for a single cell?
b) What handoff reservation scheme do the Microsurf operators have to implement for multi-cell networks in order to meet the blocking & dropping requirements above?

5.7 In a cellular system with hexagonal cells mobiles perform handover according to a simple signal-level based strategy. The path loss is distance dependent (4th-power of distance) and exhibits log-normal shadow

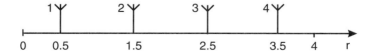

Figure 5.13 Problem 5.5.

fading (σ = 8 dB) Two cases related to different filtering strategies can be considered:

I.   Fast (instantaneous) handoff: The mobiles are always connected to the strongest base station.

II. Slow handoff: The mobiles are connected to the base station with the highest (long-term) average signal power (i.e. closest base station).

Assume that the traffic load is 50% and that the probability of nonsuccessful handoff is neglectable (enough channels for handoff). Consider sector-antennas and 1/3 and 3/9 reuse respectively (neglect the noise).

a)   Determine the outage probability (C/I-distribution) for each of the two strategies.

b)   What is the average distance to the serving base station in strategy I and II?

c)   What is the probability that the closest base station is also the one selected by strategy I?

References

[1]   Gudmundson, M., "Analysis of Handover Algorithms," *Proc. IEEE Veh. Tech. Conf., VTC '91,* St. Louis, MO, May 1991.

[2]   Gudmundson, M., "A Correlation Model for Shadow Fading in Mobile Radio," *Electronics Letters,* Vol. 27, No. 23, Nov. 1991, pp. 2146–2147.

[3]   Gudmundson, B., and O. Grimlund, "Handoff in Microcellular Based Personal Communication Systems," *Proc. WINLAB Workshop,* New Brunswick, NJ, 1990.

[4]   ETSI GSM Technical Specification 05.08—Radio subsystem link control.

[5]   Foscini, G., and M. Miljanec, "The Cost of Mobility," *IEEE Trans. Veh. Tech.*

[6]   Mouly, M., and M.-B. Pautet, *The GSM System for Mobile Communications,* published by the authors, 1992.

[7]   Tekinah, S., and B. Jabbari, "Handover Policies and Channel Assignment Strategies in Mobile Cellular Networks," *IEEE Comm. Mag.,* Vol. 29, No. 11, 1991.

[8]   Sampath, A., and J. Holtzman, "Estimation of Maximum Doppler Frequency for Handoff Decisions," *Proc. IEEE Veh. Tech. Conf., VTC '93,* Secaucus, NJ, May 1993.

[9]   Sampath, A., and J. Holtzman, "Adaptive Handoffs Through Estimation of Fading Parameters," *Proc. IEEE ICC '94,* 1994.

[10] Noerpel, A., and Y.-B. Lin, "Handover Management for a PCS Network," *IEEE Personal Communications,* Dec. 1997.

[11] Östling, P.-E., "Implications of Cell Planning on Handoff Performance in Manhattan Environments," *Int. Symp. on Personal, Indoor and Mobile Radio Comm., PIMRC,* The Hague, The Netherlands, Sept. 1994.

[12]   Markoulidakis, J. G., and E. D. Sykas, "Model for Location Updating and Handover Rate Estimation in Mobile Telecommunications," *Electronics Letters*, Vol. 29, No. 17, Aug. 1993.

[13]   Morales-Andes, G., and Villen-Altamirano, "An Approach to Modelling Subscriber Mobilty in Cellular Radio Networks," *Forum Telecom '87*, Geneva, 1987.

[14]   Kleinrock, L., *Queueing Systems, Part I: Theory*, New York: John Wiley & Sons, 1976.

6

Transmitter Power Control

6.1 Introduction

An important quantity neglected so far, is the transmitter power that the terminals and access ports should use. The transmitter power affects the link signal quality and the interference environment in a wireless system. However, adjusting the transmitter power to improve the link performance is not a trivial problem. If a terminal with a low SIR increases its transmitter power, the SIR is momentarily increased. The increase in transmitter power will on the other hand also increase the interference in the other links in the system, causing these terminals to increase their powers (power competition). Lowering the transmitter power will decrease the interference to the other links, but could, of course, jeopardize its own link.

To resolve this problem, assume an orthogonal and static channel allocation and consider the communication links that are established on an arbitrary channel, say c_0. For the uplink case, the transmitters are terminals and the receivers are the corresponding access ports. For the downlink case, their roles are reversed. Consider an instant in time in which the link gain between every receiver i and every transmitter j is stationary and is given by g_{ij}. Without loss of generality, the links are numbered such that transmitter i is the one communicating with receiver i. In a DS-CDMA system, many terminals will communicate with the same access port through a common frequency channel. In this situation, access ports i and j may denote the same physical one if the terminals i and j are assigned to the same access port.

Furthermore, assume that there are Q transmitters assigned to the channel c_0, where transmitter j uses a transmission power p_j. Let us use the following vector notation to describe all transmission powers of the transmitters:

$$P = (p_1, p_2, \ldots, p_Q)^T$$

In the uplink case, the value p_j means the transmission power of terminal j. However, in the downlink, it denotes the transmission power dedicated to terminal j by the access port to which terminal j is connected.

Equation (4.18) provides an expression for the signal to interference ratio (SIR). In other words, the interference summation is taken over all transmitters on channel c_0. If this expression is modified to take a variable transmitter power and the receiver noise (mostly thermal noise) into account, the following expression for the SIR in the receiver i on the channel can be derived:

$$\Gamma_i = \frac{g_{ii} p_i}{\displaystyle\sum_{j=1, j \neq i}^{Q} g_{ij} p_j + n_i} \tag{6.1}$$

where n_i denotes the receiver noise power at the receiver i.

Definition 6.1

The transmitter i is said to be supported if it has the SIR satisfying

$$\Gamma_i \geq \gamma_0 \tag{6.2}$$

where γ_0 is a target threshold.
Replacing (6.1) in (6.2),

$$p_i \geq \gamma_0 \left(\sum_{j=1, j \neq i}^{Q} \frac{g_{ij}}{g_{ii}} p_j + \frac{n_i}{g_{ii}} \right) \tag{6.3}$$

Equation 6.3 shows the minimal power that the transmitter i should use to achieve the target SIR, assuming the other transmitters' powers are fixed.

Example 6.1 Two-user case

Consider two terminals that are assigned to two access ports (one of each) with the channel c_0. The uplink of the system is denoted by the link gain matrix G:

$$G = \begin{pmatrix} 0.3288 & 0.0534 \\ 0.0602 & 0.3826 \end{pmatrix}$$

The target SIR γ_0 and the receiver noise are 6 dB and 0.1W, respectively. What are the power values that both terminals should use?

Solution:

With (6.3), the system can be represented by the two-dimensional graph as in Figure 6.1. In the figure, when both terminals' transmission powers belong to the region A, then both will be supported. On the other hand, when the power values are either in B or C, only terminal 1 or terminal 2 can be

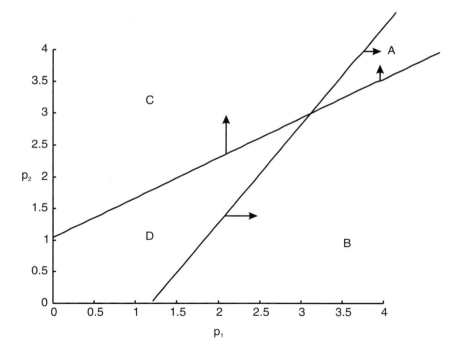

Figure 6.1 Two-user case.

supported. The worst case can be found in the region D, in which no terminal is supported. The most trivial power vector in the region A is the one that satisfies the simultaneous equation system corresponding to (6.3). In this example, the optimal point is $(p_1, p_2) = (3.1657, 3.0235)$ that minimizes the sum of transmission powers of the two terminals.

From the above example, an important question was how to maximize the number of supported terminals by selecting transmission powers appropriately. More precisely, the goal of power control is to find a power vector such that the number of supported users is maximized. In general, it is difficult to answer the question. In particular when region A does not exit (in the case when not all the terminals can be supported), it becomes a hard problem to maximize the number of supported terminals.

To understand the power control problem further, the following notations are introduced. Define the $Q \times Q$ (non-negative) normalized link gain matrix $H_{ij} = (h_{ij})$ such that $h_{ij} = \gamma_0 \dfrac{g_{ij}}{g_{ii}}$ for $i \neq j$ and $h_{ij} = 0$ for $i = j$, and the normalized noise vector $\eta = (\eta_i)$ such that $\eta_i = \gamma_0 \dfrac{n_i}{g_{ii}}$. Then, the linear inequality given in (6.3) can be rewritten to

$$p_i \geq \sum_{j=1}^{Q} (h_{ij} p_j + \eta_i)$$

With the matrix notation, the Q linear inequalities ($\Gamma_i \geq \gamma_0$, for all i) are described by

$$(I - H)P \geq \eta \tag{6.4}$$

where I denotes the identity matrix and the notation $A \leq B$ means that all components of A are less than or equal to those of B, componentwise. In (6.4), the power vector P should be non-negative for which the following proposition can be derived.

Definition 6.2

The target SIR γ_0 is said to be achievable if there exists a non-negative power vector P such that $\Gamma_i \geq \gamma_0$ for all i.

In other words, if the system of linear inequality (6.4) has a non-negative power solution, the target SIR γ_0 is achievable. For this, the following proposition can be derived from the well-known linear algebra properties:

Proposition 6.1

The target SIR γ_0 is achievable if the dominant (largest) eigenvalue of the matrix H, denoted by $\rho(H)$, is less than or equal to one. The case of $\rho(H) = 1$ will make γ_0 achievable only when the receiver noise is zero.

Proof:

From Figure 6.1, it can be easily deduced that γ_0 is achievable only when the power vector P^* that satisfies

$$(I - H)P^* = \eta \tag{6.5}$$

is non-negative. Furthermore, if $\rho(H) < 1$, then from Theorem 3.9 in [1], the inverse matrix $(I - H)^{-1} = \sum_{k=0}^{\infty} H^k$ exits and $(I - H)^{-1} > 0$. Therefore, $P^* = (I - H)^{-1}\eta \geq 0$. When $\rho(H) = 1$, the matrix $(I - H)$ becomes singular and thus only the case of $\eta = 0$ is valid with multiple non-negative solutions to (6.5).

It can easily be seen that $\rho(H)$ increases when any entry of H increases. In other words, as either the target SIR γ_0 or link gain (path loss) g_{ij} increases, it becomes difficult to achieve the target SIR γ_0. Figure 6.2 illustrates Proposition 6.1 by a two-user example. In the figure, the target SIR increases as it moves from (a) to (c).

It can be observed that the cases (a) and (d) can support both users whereas the cases (b) and (c) can only support one user at most. As mentioned earlier in this section, maximizing the number of the supported users becomes an extremely difficult problem when it comes to the cases like (b) and (c). More details about this issue will be explained in Section 6.4. For the time being, however, the discussion is confined to the cases like (a) and (d).

6.2 SIR Balancing

Consider a power control that maximizes the minimum SIR in all links. It can be proved that such a power control is achieved by making every transmitter's received SIR balanced (equalized) while keeping the balanced SIR as high as possible. To investigate this further, consider the noiseless case, $\eta = 0$. Then from (6.4), the following linear inequality system can be formulated for a given target SIR γ_0:

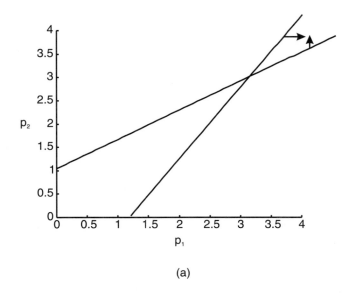

(a)

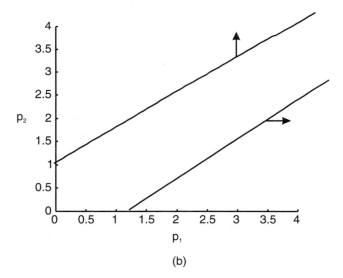

(b)

Figure 6.2 Two-user case: (a) $\rho(H) < 1$ and $\eta > 0$; (b) $\rho(H) = 1$ and $\eta > 0$; (c) $\rho(H) > 1$ and $\eta > 0$; (d) $\rho(H) = 1$ and $\eta = 0$.

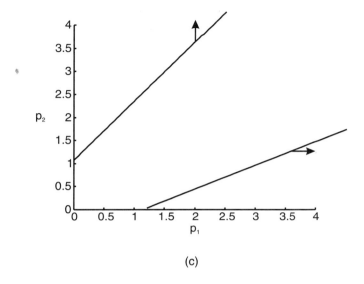

(c)

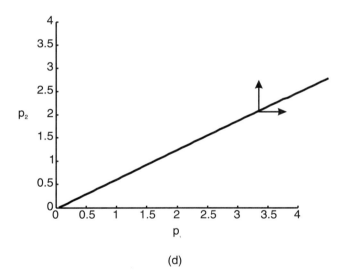

(d)

Figure 6.2 (continued).

$$(I - H)P \geq 0$$

Then, the question is to find the maximum achievable γ_0. To answer this, introduce a matrix A such that $H = \gamma_0 A$.

Proposition 6.2[1]

The inequality

$$(I - \gamma_0 A)P \geq 0 \tag{6.6}$$

has solutions in $P \geq 0$ if and only if

$$\gamma_0 \leq \frac{1}{\rho(A)} = \gamma^* \tag{6.7}$$

where $\rho(A)$ is the dominant eigenvalue of the matrix A. The power vector satisfying the expression (6.6) with equality and achieving the largest SIR γ^* for all links, is P^*, that is, the eigenvector corresponding to the eigenvalue $\rho(A)$.

Proof:

From the definition of A, it is easy to that $\rho(H) = \gamma_0 \cdot \rho(A)$ and from Proposition 6.1, $\rho(H) = \gamma_0 \cdot \rho(A) \leq 1$ should be satisfied to support every transmitter. This will lead to (6.7) (see also Figure 6.3). It is obvious that the eigenvector P^* corresponding to the eigenvalue $\rho(A)$ solves $(I - \gamma^*A)P^* = 0$ since $AP^* = \rho(A)P^*$ and $\gamma^* = \frac{1}{\rho(A)}$.

The power vector P^* will provide all users exactly with the SIR γ^*. Note that if P^* supports every user with the SIR γ^*, then the power vector cP^* also does for any scalar $c > 0$.

When the receiver noise is positive, Proposition 6.1 says that $\rho(H) = \gamma_0 \cdot \rho(A) < 1$ should be satisfied to support every user. Therefore, $\gamma^* = \frac{1}{\rho(A)}$ becomes an upper bound on the maximal achievable balanced SIR level.

---

1. This is the same as Proposition 1 in [2].

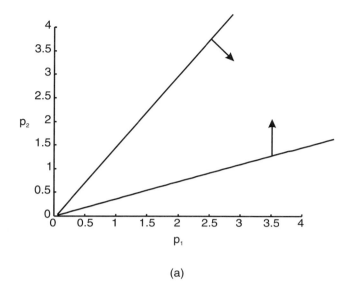

(a)

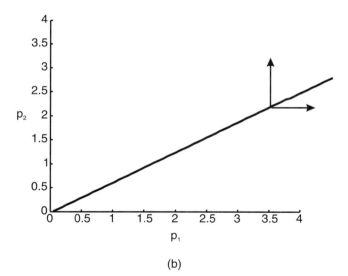

(b)

Figure 6.3 Two-user case; (a) $\rho(H) = \gamma_0 \cdot \rho(A) < 1$; (b) $\rho(H) = \gamma_0 \cdot \rho(A) = 1$.

Example 6.2 SIR Balancing

Assume that two terminals and two access ports are geographically apart as illustrated in Figure 4.1, where $r_1 = 1$ km, $r_1 = 2$ km and $D_{12} = 3$. The propagation loss is distance dependent and increases with the 4th power of the distance. Consider the uplink in the system when the receiver noise can be neglected.

   a) What is the largest minimum SIR that can be achieved in both access ports. What are the transmitter powers the terminals should use to reach this SIR level?

   b) What is the SIR in the access ports if the terminals use a constant transmitter power?

   c) What is the SIR in the access ports if the terminals control their powers to reach a constant received power at their own access ports?

Solution:

   a) First determine the link gain matrix G

$$G = \begin{pmatrix} 1^4 & 1/3^4 \\ 1/4^4 & 1/2^4 \end{pmatrix}$$

Then, the maxtrix A is given by

$$A = \begin{pmatrix} 0 & 1/3^4 \\ 1/2^4 & 0 \end{pmatrix}$$

The characteristic equation of A yields

$$\lambda^2 - 1/6^4 = 0$$

The solutions are

$$\lambda = \pm 1/6^2$$

The dominant eigenvalue $\rho(A)$ is the larger one of these two values, which according to the Proposition 6.2, corresponds to an SIR

$$\Gamma_1 = \Gamma_2 = \gamma^* = 6^2 \approx 15.6 \text{ dB}$$

This result is obtained when the power vector P^* is chosen as an eigenvector corresponding to $\rho(A)$:

$$P^* = c(1, 2.25)$$

where user 2 uses more than twice the power than user 1.

b) Let $P = (1, 1)$. From (6.1),

$$\Gamma_1 \approx 81 \ (19 \text{ dB}), \ \Gamma_2 \approx 6 \ (12 \text{ dB})$$

Note that the minimum SIR in this case is only 12 dB as compared to 15.6 dB in the previous case.

c) To achieve a constant received power at the access port, choose the following power vector for the terminals

$$P = c\left(\frac{1}{g_{11}}, \frac{1}{g_{22}}\right) = c'\left(1, \frac{g_{11}}{g_{22}}\right) = c'(1, 2^4)$$

Compute the SIR as

$$\Gamma_1 \approx 5 \ (7 \text{ dB}), \ \Gamma_2 \approx 256 \ (24 \text{ dB})$$

Note that the minimum SIR is even less in this case than in the case b. Constant received power control may thus make the SIR situation worse than using no power control at all.

If the maximum achievable SIR γ^* is larger than the given target SIR γ_0, all links reach acceptable performance. On the other hand, if γ^* is smaller than γ_0, SIR balancing would be disastrous, since all links would drop below the threshold and become useless. In the latter case there are more connections than can be supported by the channels (see Figure 6.4).

For the latter case, methods have been devised to systematically remove terminals with the objective to maximize the number of connections with sufficient SIR. It can be shown that in the optimum power allocation that maximizes the number of supported users, some links are completely shut off, and that all remaining supported users have SIR-balanced links. For instance, consider the following removal algorithm given in [2].

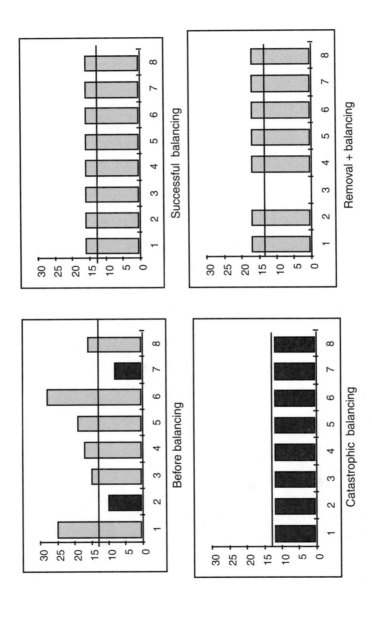

Figure 6.4 SIR Balancing—eight-user case.

Algorithm SRA (Stepwise Removal Algorithm)

Step 1. Determine γ^* corresponding to the matrix A. If $\gamma^* \geq \gamma_0$ use the eigenvector P^*, else set $Q' = Q$.

Step 2. Remove the temonal k, for which the maximum of the row and column sums (in the matrix A)

$$r_k = \sum_{j=1}^{Q} a_{kj}, \; c_k = \sum_{j=1}^{Q} a_{jk}$$

is maximized and form $(Q' - 1) \times (Q' - 1)$ matrix A'. Determine γ^* corresponding to the matrix A'. If $\gamma^* \geq \gamma_0$ use the eigenvector P^*, otherwise set $Q' = Q' - 1$ and repeat Step 2.

The idea in SRA is to remove one user at the time until the required SIR level is achieved in the remaining links. Some typical performance results from SRA are given in Figure 6.5. As the graph shows, the gain in capacity (cluster size) by using the combination of SIR-balancing and removals is in the order of 300–400%.

6.3 Distributed Power Control

6.3.1 Iterative Method

One assumption used so far is that the link gain matrix G is known. This assumption, however, is impractical since it requires a centralized measurement mechanism for the system, which in turn will require very heavy signaling between access ports and terminals. In this section, focus is on how to avoid such a centralized control and how to design distributed power control algorithms.

Consider the power control problem given in (6.4). Since the ideal situation is to make connection with the minimal transmission power, the following equality constraint on SIR can be considered:

$$(I - H)P = \eta \tag{6.8}$$

Throughout Sections 6.3.1 to 6.3.5, it is assumed that the receiver noise is not negligible and that there exists a unique and non-negative power vector P^* that solves the problem (6.8). In other words, $\rho(H) < 1$ so that the matrix $(I - H)$ is nonsingular and $P^* = (I - H)^{-1}\eta \geq 0$.

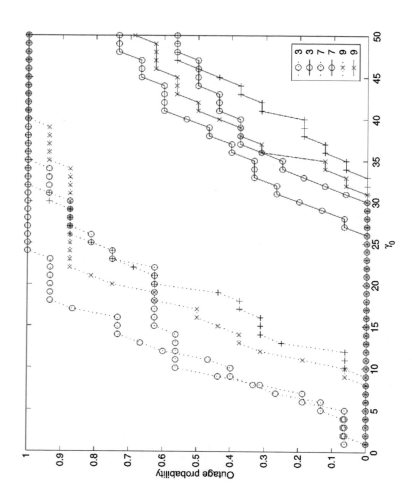

Figure 6.5 Outage probability as a function of the required SIR for a system with static channels for $K = 3$, 7, and 9; Lognormal fading, $\sigma = 6$ dB; Upper curves, constant transmitter power; Lower curves, SRA power control.

Since every element in the matrix H is hardly available in practical systems, efficient methods, that is, Gaussian elimination for solving the linear equation system (6.8) cannot be utilized. Only those iterative methods that can be executed with local measurement and signaling have drawn much attention.

The theoretical roots of many iterative power control algorithms can be found in numerical linear algebra. Consider the following general iterative method for solving (6.8):

$$P^{(n+1)} = M^{-1}NP^{(n)} + M^{-1}\eta, \ n = 0, 1, \ldots \tag{6.9}$$

where M and N are matrixes of appropriate sizes such that

$$P^* = M^{-1}NP^* + M^{-1}\eta$$

The vector $P^{(n)}$ denotes the power level at iteration n. Appropriately selecting M and N, the above iterative method can converge, that is,

$$\lim_{n \to \infty} P^{(n)} = P^* = (1 - H)^{-1}\eta$$

By defining $M = I$ and $N = H$, a power control algorithm can be constructed from (6.9):

$$P^{(n+1)} = HP^{(n)} + \eta, \ n = 0, 1, \ldots$$

It is easy to see that, for each transmitter i, the method becomes

$$p_i^{(n+1)} = \frac{\gamma_0}{g_{ii}}\left(\sum_{j=1, j \neq i}^{Q} g_{ij}p_j^{(n)} + n_i \right) = \frac{\gamma_0}{\gamma_i^{(n)}} p^{(n)}, \ n = 0, 1, \ldots$$

where $\gamma_i^{(n)}$ and $p_i^{(n)}$ denote the received SIR and transmission power of transmitter i at iteration n, respectively. This algorithm is denoted by DPC (Distributed Power Control). Note that DPC requires only local measurement on SIR at the receivers. DPC was suggested by Foschini and Miljanic [3], which is the same as the Jacobi relaxation method used in the numerical linear algebra [1].

6.3.2   Convergence of the Iterative Method

In the general iterative method (6.9), let α_1, α_2, be the eigenvalues of the iteration matrix, $M^{-1}N$ and define $\rho(M^{-1}N) = \max_k |\alpha_k|$. Furthermore, the error vector is defined by

$$\epsilon^{(n)} = P^{(n)} - P^*$$

From (6.9), the error vector $\epsilon^{(n)}$ can be expressed as

$$\epsilon^{(n)} = M^{-1}N \cdot \epsilon^{(n-1)} = \ldots = (M^{-1}N)^n \cdot \epsilon^{(0)} \qquad (6.10)$$

Then, in order for (6.10) to converge to zero vector, it can be proved that $\rho(M^{-1}N) < 1$ should hold. This is summarized in the following proposition.

Proposition 6.3 [1]

For an achievable target SIR γ_0, the error vector $\epsilon^{(n)}$ tends to the zero vector starting with any $\epsilon^{(0)}$ if and only if $\rho(M^{-1}N) < 1$.
From (6.10), it can be observed that the power vector converges to a fixed point with a geometric rate whenever it converges. In the case of DPC, $\rho(M^{-1}N) = \rho(H)$ and it is known from Proposition 6.1 that $\rho(H) < 1$ when the target SIR is achievable and receiver noise is positive. Therefore DCP will converge to P^* whenever the given target is achievable.

6.3.3   Convergence Speed of the Iterative Method

Along with distribution, convergence speed of power control is another important criterion by which one can determine the practical applicability of a given power control algorithm. This argument is quite reasonable since it has been assumed so far that the link gain matrix is invariable during the power control process. However, the values of the gain matrix as well as the size of the matrix are changing continuously due to mobile movement and the propagation condition change. Thus, a good power control algorithm should quickly and distributively converge to the state where the system supports as many users as possible. It was shown (Theorem 3.2 in [1]) that the smaller $\rho(M^{-1}N)$ gives the faster convergence. Therefore, an interesting task is to find M and N that give $\rho(M^{-1}N) < 1$ as small as possible. At the same time such matrixes should lead to the distributed power update form as in DPC.

Example 6.3 Convergence Speed of DPC

In the two-user example given in Example 6.1, consider the DPC algorithm and a power control algorithm (called Unconstrained Second-Order Power Control (USOPC) [4]) given by

$$p_i^{(n+1)} = \omega \frac{\gamma_0}{\gamma_i^{(n)}} p_i^{(n)} + (1 - \omega) p_i^{(n-1)}, \ n = 1, 2, \ldots$$

where $p_i^{(1)} = \frac{\gamma_0}{\gamma_i^{(0)}} p_i^{(0)}$ and $\omega = 1.1291$. Assume that the initial power vector is given by $P^{(0)} = (1.8772, 0.521)$. How many iterations (power updates) are needed for DPC and USOPC to support both terminals so that the normalized Euclidean error $\| \epsilon^{(n)} \| / \| \epsilon^{(0)} \|$ is less than 0.0001?

Solution

The power vectors generated from DPC and USPC are depicted in Figures 6.6 and 6.7. It takes 11 iterations for USOPC, whereas for DPC, it takes more than 20 iterations to reach the state of the normalized Euclidean error $\| \epsilon^{(n)} \| / \| \epsilon^{(0)} \|$ less than 0.0001.

In Example 6.3, USOPC can be proved to be faster than DPC, when ω is chosen carefully from the open interval $(1,2)$ (see Section III in [4]). The power update form of USOPC can be rewritten as

$$p_i^{(n+1)} = \frac{\gamma_0}{\gamma_i^{(n)}} p_i^{(n)} + (\omega - 1) \left(\frac{\gamma_0}{\gamma_i^{(n)}} p_i^{(n)} - p_i^{(n-1)} \right), \ n = 1, 2, \ldots$$

In calculating $p_i^{(n+1)}$, the algorithm first compares the result from DPC, $\frac{\gamma_0}{\gamma_i^{(n)}} p_i^{(n)}$ with the previous power value $p_i^{(n-1)}$. If $\frac{\gamma_0}{\gamma_i^{(n)}} p_i^{(n)} > p_i^{(n-1)}$, the algorithm tries to choose a power value $p_i^{(n+1)}$ such that $p_i^{(n+1)} > \frac{\gamma_0}{\gamma_i^{(n)}} p_i^{(n)}$. The gap between $p_i^{(n+1)}$ and $\frac{\gamma_0}{\gamma_i^{(n)}} p_i^{(n)}$ becomes smaller as either $\omega - 1$ or $\frac{\gamma_0}{\gamma_i^{(n)}} p_i^{(n)} - p_i^{(n-1)}$ is approaching zero. With the

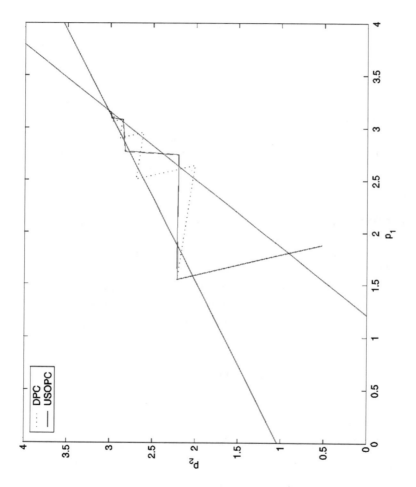

Figure 6.6 Traces of DPC and USOPC.

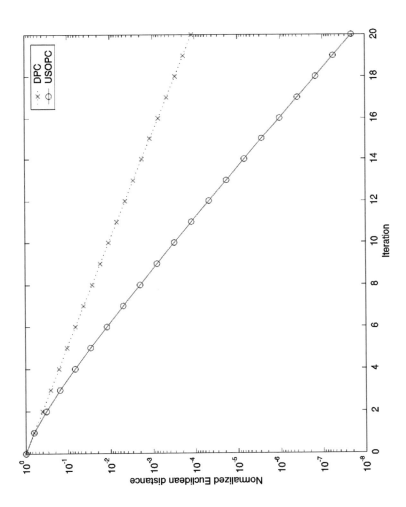

Figure 6.7 Normalized Euclidean distance between the current power vector and P^* as a function of iteration ($\|\epsilon^{(n)}\|/\|\epsilon^{(0)}\|$).

same reasoning, it can be seen that $p_i^{(n+1)} < \dfrac{\gamma_0}{\gamma_i^{(n)}} p_i^{(n)}$ if

$\dfrac{\gamma_0}{\gamma_i^{(n)}} p_i^{(n)} < p_i^{(n-1)}$. When $\dfrac{\gamma_0}{\gamma_i^{(n)}} p_i^{(n)} = p_i^{(n-1)}$, it is easy to see

$p_i^{(n+1)} = \dfrac{\gamma_0}{\gamma_i^{(n)}} p_i^{(n)}$ (same as DPC).

6.3.4 Standard Interference Function

Due to the framework given in [5], the convergence of the distributed (and iterative) power control algorithms can be proved from a slightly different angle. The power control algorithms that have an iterative nature, can be described by the following general function:

$$P^{(n+1)} = I(P^{(n)})$$

where the function I, called the *interference function*, will define the power vector of the next iteration.

Definition 6.3

Assuming positive receiver noise, an interference function I is called standard if it satisfies for all non-negative power vectors:

$$I(P) > 0$$

$$P \geq P' \Rightarrow I(P) \geq I(P')$$

$$\forall \alpha > 1, \; \alpha \cdot I(P) > I(\alpha P)$$

For the standard interference function, the following theoretic property can be proved.

Proposition 6.4 (Theorem 2 in [5])

The sequence of power vectors from the standard interference function will converge to the solution of (6.8), starting with any non-negative power vector, when the system (6.8) has the unique non-negative solution P^*.

It is easy to show that DPC belongs to the class of standard interference function. From the power update form of DPC,

$$p_i^{(n+1)} = \frac{\gamma_0}{g_{ii}} \left(\sum_{j=1, j\neq i}^{Q} g_{ij} p_j^{(n)} + n_i \right) = \frac{\gamma_0}{\gamma_i^{(n)}} p_i^{(n)}$$

The first two conditions are satisfied obviously. Since

$$\alpha p_i^{(n+1)} = \alpha \frac{\gamma_0}{g_{ii}} \left(\sum_{j=1, j\neq i}^{Q} g_{ij} p_j^{(n)} + n_i \right) > \frac{\gamma_0}{g_{ii}} \left(\alpha \sum_{j=1, j\neq i}^{Q} g_{ij} p_j^{(n)} + n_i \right)$$

the last condition also holds.

However, being a standard interference function is a sufficient condition to the convergent power control. Convergent power control algorithms that do not belong to the class of standard interference functions can be constructed. For instance, USOPC in Example 6.3 does not fulfill the above condition. In particular, the first condition can be violated when $\frac{\gamma_0}{\gamma_i^{(n)}} p_i^{(n)} < p_i^{(n-1)}$. However, its convergence can be proved using Proposition 6.3 (see [4] for more details).

6.3.5 Constrained Power Control

It has been assumed that the transmitter power can be adjusted without limitations. This is not realistic, since the maximum output power of a transmitter is upper-bounded in any implementation. In particular for hand-held terminals, this constraint is critical, mainly due to the battery capacity. To consider this limitation in power control, let's introduce a constraint given by:

$$0 \leq P \leq \overline{P} \tag{6.11}$$

where $\overline{P} = (\overline{p}_1, \overline{p}_2, \ldots, \overline{p}_Q)$ denotes the maximum transmission power of each transmitter. If the required power is greater than the transmitter's maximum peak power level, some action has to be executed to constrain the power value within the power range. For example, DPC in Section 6.1 can be modified into the power-constrained version:

$$p_i^{(n+1)} = \min \left\{ \frac{\gamma_0}{\gamma_i^{(n)}} p_i^{(n)}, \overline{p}_i \right\}, \quad n = 0, 1, \ldots$$

This algorithm is called *Distributed Constrained Power Control* (DCPC) [6]. In this algorithm, when the required power from DPC power update is greater than the maximum power, the transmitter uses the maximum power.

DCPC has a property that the power may reach the maximum level when a user is experiencing low channel quality. Unfortunately, even if the maximum power is used, this may not necessarily lead to sufficient improvement on channel quality in particular when the user is located in the unfavorable position. The impact will be high power consumption and severe interference, hitting other users. With this in mind (making DCPC a special case), a more general scheme can be described by

$$
p_i^{(n+1)} = \begin{cases} \dfrac{\gamma_i}{\gamma_i^{(n)}} p_i^{(n)} & \text{if, } \dfrac{\gamma_i}{\gamma_i^{(n)}} p_i^{(n)} \le \overline{p}_i \\[2ex] \hat{p}_i & \text{if, } \dfrac{\gamma_i}{\gamma_i^{(n)}} p_i^{(n)} > \overline{p}_i \end{cases} \tag{6.12}
$$

where $0 \le \hat{p}_i \le \overline{p}_i$.

Proposition 6.5[2]

The constrained power control algorithm (6.12) will converge to the solution of (6.8) starting with any non-negative power vector, when the system (6.8) has the unique solution P* within the power range (6.11).

When the channel quality is poor, it is not necessary to use the maximum power. Under poor conditions, the power may even be lowered to the minimum level. In that case the user stays on the same channel and transmission will be resumed, if the interference situation becomes favorable. The main motivation of this power control is to save energy consumption by not transmitting the maximum power when the transmitter cannot be supported within its power range.

Note that DCPC belongs to the class of standard interference function. However, the general constrained power control algorithm (6.12) does not fulfil the second condition of the standard interference function.

6.3.6   Distributed SIR Balancing

The SIR balancing technique discussed in Section 6.2 is a centralized scheme, requiring full knowledge of the link gain matrix. A distributed SIR balancing algorithm was also suggested in [8] for the noiseless system case:

---

2. This is the same as Proposition 3 in [7].

Algorithm DB (Distributed Balancing Algorithm)

$$p_i^{(n+1)} = \beta \cdot p_i^{(n)} \left(1 + \frac{1}{\gamma_i^{(n)}}\right), \quad n = 0, 1, \ldots$$

where β is any positive constant.

It was shown in [8] that $\lim_{n \to \infty} P^{(n)} = P^*$ and $\lim_{n \to \infty} \gamma^{(n)} = \gamma^*$ (starting with an arbitrary positive power vector), where γ^* is the maximum achievable SIR level and P^* is the corresponding power vector defined in Proposition 6.2.

A practical problem is that the transmitter powers in the DB algorithms are all increasing, unless the parameter β is chosen properly. Since in SIR balancing, the transmission power ratio among the terminals is more important than the actual power values, selecting

$$\beta = \beta^{(n)} = \frac{1}{\displaystyle\sum_{i=1}^{Q} p_i^{(n)}}, \quad n = 1, 2, \ldots$$

would ensure a constant sum of powers of all terminals. However, calculating this quantity may not be possible in a completely distributed way.

6.4 Transmitter Removal

Consider Example 6.1. What will happen if the maximum transmission power of the terminals is 3W (see Figure 6.8) Since the power vector that will support both users with the minimal power consumption is (3.1657, 3.0235), it is impossible to support both users simultaneously. In this example, DCPC will converge to the power vector (3,3), supporting neither of the users. This means that the system does not support any user, while each transmitter consumes much energy. This kind of situation is often called the cocktail effect especially in DS-CDMA systems.

So far it has been assumed that the given target SIR is achievable. What will happen if the target is not achievable? Further, what will happen if the target SIR is achieved only by the power vectors that are in the outside of the power range? In either case, distributed power control algorithms described here will behave quite undesirably. In general a target SIR becomes unachievable when an overload situation occurs on a channel. In this con-

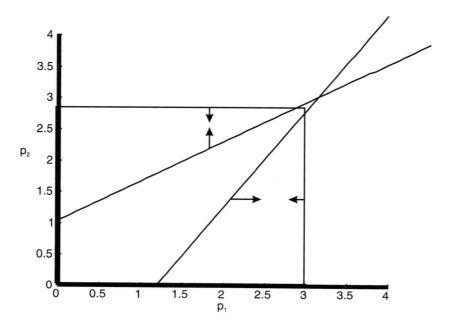

Figure 6.8 Two-user example: Constrained power case.

gested situation, some of the users should be removed (i.e., powered off). By doing this, the remaining users can be supported with reduced power consumption. The removed users are either handed over to another channel or disconnected, which is highly undesirable for the radio network and subscribers. Therefore, an interesting problem is to minimize the number of removed users. This problem unfortunately belongs to the class of NP-complete optimization problems [9]. Getting an optimal solution will require computation effort that grows exponentially according to the number of active users in the system. Therefore, previous results on the transmitter removal are mostly restricted to efficient heuristic algorithms.

6.5 Dynamic Behavior of Power Control

For analytic simplicity, the so-called *snapshot* approach has been used in this chapter. More precisely, the link gain matrix is assumed to be fixed during power control. In this section, this assumption is relaxed to see the dynamic behavior of power control. For the purpose, DCPC described in Section

6.3.5 is implemented using the functions provided by RUNE in Appendix B.

The considered system is composed of 27 regular hexagonal cells with the cluster size $K = 3$ (total 9 clusters) and the cell radius 1 km. Terminals are moving around the service area with mean speeds, 0, 5, 20 m/s respectively. The DCPC power update is done at the terminals in every second with the target SIR 10dB. Maximum peak transmission power of the terminal is 3 dBm. The receiver noise at base stations are set to −118 dBm. The distance attenuation coefficient is 4 and the standard deviation for the log-normal fading is set to 6 dB. One typical terminal is picked (and traced). RUNE simulation is done until the terminal finishes 50 power updates (50 seconds). Figure 6.9 shows the received SIR of the terminal at the base station at each power update. For the stationary case (mean speed = 0), the target SIR 10 dB is reached in less than 20 power updates. However, as the terminal speed increases, the received SIR is fluctuating around the target. This means that the propagation condition (link gain matrix) is changing before DCPC converges to the target SIR.

There are basically two approaches to resolve such a large SIR-fluctuation. First, the power update interval can be decreased, for instance, hundreds of times per second. This solution has, of course, a drawback of adding a burden of frequent measurement and signalling to the system. The second but more fundamental approach is to develop a faster power control algorithm than DCPC.

As will be discussed in Section 9.6, the power control algorithms discussed so far cannot be implemented in real systems without any modification, due to some practical limitations. This makes the efficiency of power control even worse. In many cases, theoretically faster power control algorithms are not easy to implement.

6.6 Multirate Power Control

Up until now, the same target SIR is assumed for all users. However, note that all the Propositions (except SIR balancing) can be easily extended to the situation where each user has his or her own target SIR. It will be quite common in the next generation cellular system that each user has a respective target SIR. This is mainly because users will access different services with different transmission rate and error requirements and in radio channels, the transmission rates and errors are closely related to the SIR. The availability of variable transmission rates in a radio network raises the problem of

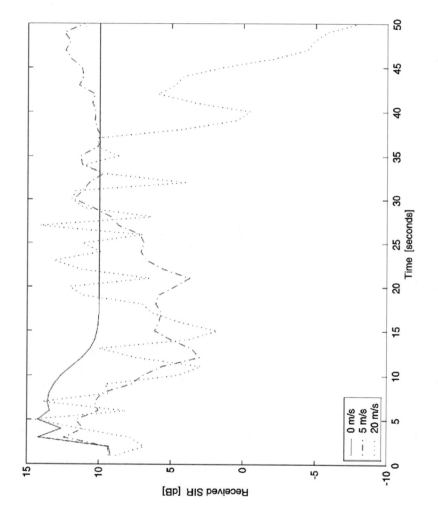

Figure 6.9 SIR of a moving terminal (DCPC power update).

controlling them in the most spectrally efficient way. In particular, since SIR is properly controlled by the power control, the combined rate and power control becomes an interesting problem. In this section, some fundamental insight into the issue is discussed.

Consider a wireless network consisting of access points and terminals. At a certain instance of time, there are a number of active links 1, 2, 3 ... Q. For each link i the emitted power of the transmitter, p_i and the link data rate r_i can be controlled. The rate adaptation mechanism allows the data rate

$$r_i \leq f(\gamma_i)$$

where γ_i is the SIR at the receiver i, and f is a monotonically increasing modem (waveform set/coding) dependent function. The SIR at the receiver i is given by

$$\gamma_i(P) = \frac{g_{ii}p_i}{\sum_{j=1, j \neq i}^{Q} g_{ij}p_j + n_i} \tag{6.12}$$

Assume that each transmitter may use the maximum transmission power $\overline{P} = (\overline{p}_1, \overline{p}_2, \ldots, \overline{p}_Q)$.

Definition 6.4

A rate vector $R(\overline{P}) = (r_1, r_2, \ldots, r_Q)$ is instantaneously achievable if there exists a positive power vector

$$P = (p_1, p_2, \ldots p_Q) \leq \overline{P}$$

such that

$$r_i \leq f(\gamma_i(P)), \forall \ i$$

Definition 6.5

A rate vector $R^*(\overline{P}) = (r_1^*, r_2^*, \ldots, r_Q^*)$ is achievable in the average sense if it may be expressed as

$$R^* = \sum_k \alpha_k R_k$$

where

$$\alpha_k \in [0, 1], \sum_k \alpha_k = 1$$

where all the R_k are instantaneously achievable rate vectors.

If a set of rate vectors R_k is instantaneously achievable, it is possible to switch between rate vectors, using each of them during the fraction of time α_k (TDM, or allocated this fraction of bandwidth to this particular set of rates, FDM), and yielding the average rate $R^*(\overline{P}) = (r_1^*, r_2^*, \ldots, r_Q^*)$.

Now, assume that each link i requires a minimum data rate, denoted $R_{i,\min}$. Furthermore, assume that any excess data rate provided to the user is potentially consumed (and paid for) by the user. The operator therefore has an interest to provide as much excess data rate as possible. Furthermore, if the connection is nonreal time traffic, the following optimization problem can be considered.

$$\max \sum_{i=1}^{Q} r_i^*(\overline{P})$$

subject to

$$r_i^*(\overline{P}) \geq r_{i,\min}, \forall i$$

Use the following two sample relations between the data rate and the SIR γ:

Case A: $f(\gamma) = c\gamma$

Case B: $f(\gamma) = c'\log(1 + \gamma)$

The Case A corresponds to the case where the available bandwidth is unlimited and the same waveform is scaled to achieve a constant E_b/N_0 to provide a constant link quality in terms of bit error probability, whereas the case B is simply the Shannon limit for a band-limited channel. Since, in both cases there is a one-to-one correspondence between the instantaneously achievable data rates and the (required) link SIR, the sets of achievable rates (SIR) can be investigated.

For two-user cases, simplifying (6.12) gets

$$
\begin{array}{rl}
p_1 \qquad -\gamma_1 a_{12} p_2 & \geq \gamma_1 \eta_1 \\
-\gamma_2 a_{12} p_2 \qquad + p_2 & \geq \gamma_2 \eta_2
\end{array}
$$

where

$$
a_{ij} = \frac{g_{ij}}{g_{ii}} \text{ and } \eta_i = \frac{n_i}{g_{ii}}
$$

and γ_i is the minimum SIR required to achieve instantaneous rate r_i in link i. Solving for p_1 and p_2 yields

$$
\begin{aligned}
0 \leq p_1 &= \frac{\gamma_1(\eta_1 + \gamma_2 a_{12} \eta_2)}{1 - \gamma_1 \gamma_2 a_{12} a_{21}} \leq \bar{p}_1 \\
0 \leq p_2 &= \frac{\gamma_2(\eta_2 + \gamma_1 a_{21} \eta_1)}{1 - \gamma_1 \gamma_2 a_{12} a_{21}} \leq \bar{p}_2
\end{aligned}
\qquad (6.13)
$$

Since both power values should be non-negative, the following necessary condition on the achievable SIR can be derived

$$
1 - \gamma_1 \gamma_2 a_{12} a_{21} > 0
$$

$$
\gamma_1 \gamma_2 < \frac{1}{a_{12} a_{21}}
$$

Now, applying the two bandwidth-quality criteria A and B to (6.13), the results illustrated in Figures 6.10 and 6.11 can be obtained. Figure 6.10 illustrates the bandwidth unlimited case. Here it can be seen that the achievable instantaneous rate regions (as indicated by the region below the solid lines) are not convex, but the average sense rate regions (as indicated by dotted lines) are. The average rate region is in fact given by a set of linear equations (straight lines). The figure also illustrates a minimum rate requirement $r_i > R_{i,\min}$ (i.e., within the rectangle in the upper right corner). In this example, it can be seen that for the maximal power 5, there is no rate pair meeting the constraints, neither instantaneously nor in the average sense. For the maximal power 10 the rate requirements can be achieved in the average sense but not instantaneously. For the highest maximal power, the rate constraints can be met in both senses. In this case the sum rate is maximized by time multiplexing individual transmissions. Since the average rate region is given by a linear expression, the maximal average sum rate is

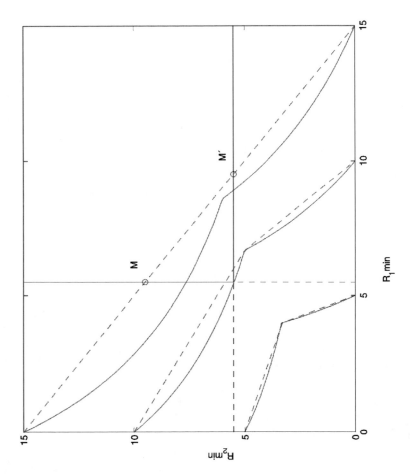

Figure 6.10 Achievable rates for various maximal transmitter powers. No bandwidth limitation (Case A). Relative maximal powers 5, 10, and 15. $a_{12} = 0.05$, $a_{21} = 0.02$, $\eta_1 = \eta_2 = 1$.

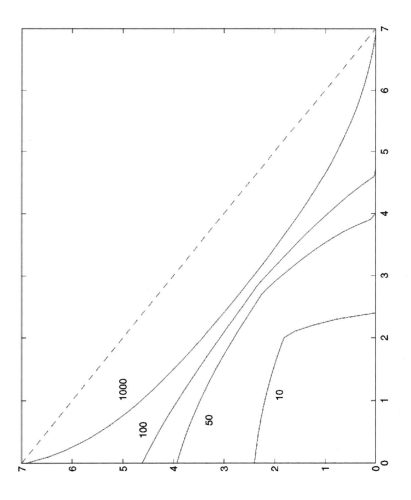

Figure 6.11 Achievable rates for various maximal transmitter powers. No bandwidth limitation (Case B). Relative maximal powers 10, 50, 100, and 1000. $a_{12} = 0.05$, $a_{21} = 0.02$, $\eta_1 = \eta_2 = 1$, $c' = 1$.

found in one of the corners of the achievable and feasible rate region, that is, in either point M or M' (M is the optimum in this particular example).

Figure 6.11 illustrates the Case B. Here, in contrast to the previous case, the instantaneous rate regions are convex for low maximal transmitter powers. This means that for these maximum powers, any achievable rate pair can be achieved instantaneously, that is, by assigning appropriate powers both transmitters may transmit simultaneously. For higher transmitter powers, the regions become convex. The higher (average) sum rates can be achieved by time multiplexing the transmissions.

The results show that the maximal total (sum) data rates that can be achieved in the average sense by single transmissions are in many cases (but not always) higher. In particular this holds true in a high power (or short range) scenario where inter-user interference is the dominating source of disturbance. The latter statement also applies to the instantaneous sum rate. For lower power levels (e.g., longer ranges), maximal sum rates are mostly reached for simultaneous transmission schemes.

Note that the characterizations are general in the sense that they are not sensitive to the choice of modulation schemes and apply for CDMA type waveforms as well. As a practical consequence, the results demonstrate that for short range, interference limited systems (e.g., indoor, microcell) coordinated orthogonal multiplexing provides better performance than simultaneous transmissions (e.g., CDMA). For longer ranges when noise (i.e., coverage) comes into play nonorthogonal schemes have advantages.

Problems

5.1 To conserve battery power in the mobile terminal, the following power control scheme is used in a wireless cellular system. Each terminal controls its power to achieve a constant received power at the intended receiver (access port). Estimate the reduction in average transmitter power in a mobile terminal, compared to a system with constant power. Use the approximation that the terminals are uniformly distributed over circular cells. The propagation loss is proportional to the distance to the power α.

5.2 A new cellular operator has installed a small mobile telephone system along a street. The system has only 6 base stations located at the following distances (coordinates), measured from one end of the street:

$$(100,200,300,400,500,600).$$

The channel allocation scheme has, at some point, assigned the channel to four mobiles at the locations

$$(50,280,420,545).$$

a) Assuming the propagation loss is proportional to the fourth power of the distance, determine the link gain matrix for this channel set.
b) All base stations and mobile terminals are assumed to use the same (constant) transmitter power. The mobiles are assumed to be connected to the base station with the lower path loss. Determine the SIRs for all base stations and mobiles on this channel.
c) In an attempt to conserve battery power and possibly to improve the SIR, the operator introduces a power control regime that maintains a constant received power at the intended receiver. Determine the transmitter powers and the resulted SIR in the base stations and mobiles under this regime.
d) Compare the results in c) with the case where a SIR-balancing power control scheme is employed.

5.3 A wireless cellular communication system with 3 cells and 2 channels has at some point 5 active terminals. The system employs SIR-balancing power control. Both channels may be used in all cells if an SIR of at least $\gamma_0 = 18$ dB is achieved. Measurements show that link gain matrix is given by

$$G = \begin{pmatrix} 2 \cdot 10^{-7} & 2 \cdot 10^{-7} & 4 \cdot 10^{-8} & 2 \cdot 10^{-9} & 2 \cdot 10^{-10} \\ 2 \cdot 10^{-9} & 1 \cdot 10^{-7} & 2 \cdot 10^{-6} & 1 \cdot 10^{-7} & 2 \cdot 10^{-9} \\ 2 \cdot 10^{-10} & 2 \cdot 10^{-9} & 4 \cdot 10^{-9} & 2 \cdot 10^{-7} & 2 \cdot 10^{-7} \end{pmatrix}$$

a) Is there a channel assignment for which the SIR requirement is met in the down channel for all terminals?
b) Study the SIRs that can be achieved in the uplink for this channel assignment. Conclusion?

5.4 Consider two terminals that are assigned to two access ports (one of each) using the same frequency channel. The uplink is denoted by the link gain matrix:

$$G = (g_{ij}) = \begin{pmatrix} 1/3 & 1/21 \\ 1/20 & 1/4 \end{pmatrix}$$

Assume that the receiver noise level is 0.1W, and answer questions below.

a) Consider a power control that tries to make the minimum received SIR greater than 7 dB. Is this possible? (neglect receiver noise).

b) Assume that the target SIR is 6 dB. What are the transmitter powers that can achieve the target SIR in both terminals with the minimal transmitted powers?

c) Assume an initial power vector $P^{(0)} = (1.8772, 0.5211)$. How many iterations (power updates) are needed for DCPC (target SIR 6dB) to support both mobiles with the received SIR of 5 dB?

d) Assume that the maximum transmission power of the terminals is 3W and answer the question b) again. Is it possible to support both terminals? If not, which terminal should be removed (power off)?

5.5 A cellular mobile communication system uses a 3/9 reuse pattern. The traffic can be assumed to correspond to an activity factor of 50%. Two power control algorithms are to be compared: Optimum SIR-balanced power control (without removals) and a partial compensation power control scheme where the transmitter power is determined by:

$$P_{\text{tx}} = P_0 / G^\beta$$

where β is a design parameter to choose. Determine which of the algorithms provides the highest SIR-threshold for 10% outage probability. Assume the system is interference limited (neglect the noise).

References

[1]   Varga, R. S., *Matrix Iterative Analysis*, Englewood Cliffs, NJ: Prentice-Hall, 1962.

[2]   Zander, J., "Performance of Optimum Transmitter Power Control in Cellular Radio Systems," *IEEE Trans. Veh. Technol.*, Vol. VT-41, 1992, pp. 57–62.

[3]   Foschini, G. J., and Z. Miljanic, "A Simple Distributed Autonomous Power Control Algorithm and Its Convergence," *IEEE Trans. Veh. Technol.*, Vol. VT-42, 1993, pp. 641–646.

[4]   Jäntti, R., and S.-L. Kim, "Second-Order Power Control with Asymptotically Fast Convergence," *IEEE Journal Sel. Areas Commun.*, Vol. SAC-18, 2000, pp. 447–457.

[5]   Yates, R., "A Framework for Uplink Power Control in Cellular Radio Systems," *IEEE J. Sel. Areas Commun.*, Vol. SAC-13, 1995, pp. 1341–1348.

[6]   Grandhi, S. A., J. Zander, and R. Yates, "Constrained Power Control," *Wireless Personal Communications*, Vol. 1, 1995, pp. 257–270.

[7]   Berggren, F., R. Jäntti, and S.-L. Kim, "A Generalized Algorithm for Constrained Power Control with Capability of Temporary Removal," submitted to *IEEE Trans. Veh. Technol.*, 1999.

[8]   Zander, J., "Distributed Cochannel Interference Control in Cellular Radio Systems," *IEEE Trans. Veh. Technol.*, Vol. VT-41, 1992, pp. 305–311.

[9]   Andersin, M., Z. Rosberg, and J. Zander, "Gradual Removals in Cellular PCS with Constrained Power Control and Noise," *ACM/Baltzer Wireless Networks Journal*, Vol. 2, 1996, pp. 27–43.

—

7

Dynamic Channel Allocation

7.1 Introduction

As has been already noted in previous chapters, both the number of active terminals as well as the signal-to-interference ratio (caused by varying propagation conditions) exhibit considerable fluctuations. Previous chapters show that considerable design margins have to be added regarding the reuse distance in order to provide a sufficiently low interference rate χ. The price paid is a low number of channels in each cell. At the same time, the number of available channels has to be sufficiently large to avoid temporary overloading, in order to keep the primitive assignment failure rate ν_p at bay. Since these two goals are somewhat in contradiction with each other, a compromise has to be reached. Conventionally, static channel allocation schemes (FCA, fixed channel allocation) are conservatively designed and characterised by rather low interference rates. On the other hand, there is a rather low resource utilization, since only a few channels are used ($q \ll 1$). The latter problem is particularly highlighted if the number of available channels per cell η is low (due to trunking losses). Further, there are occasions, when, due to fluctuation in traffic, calls are blocked, even if there are channels available in adjacent cells.

The resulting poor spectrum utilization has to be attributed to the fact that knowledge about actual, instantaneous traffic situations and the propagation conditions, at the time of the allocation of resources is limited. More detailed information would make it possible to predict the number of terminals, the interference level, and so forth with a lower variance. This is done in the planning and installation stage of most contemporary mobile telephone systems. The static channel allocation is often preceded by measurements and predictions of the traffic load and the propagation conditions. Modern propagation prediction tools based on terrain databases and using diffraction/reflection models are used in the planning process of all wide-area covering mobile telephone systems. The result may be an irregular cell plan in which the distance between two cells using the same channel group may depend on the geometry of the terrain. Further, adapting to the (average) traffic load may cause different cells (RAPs) to be assigned a different number of channels. The result is a much better resource utilization and a higher capacity.

Detailed planning improves the performance of the system but has its limitations. What one can hope to predict, is the average traffic load and the local average propagation conditions. Since terminals become active in a random fashion and move about the service area in a way that may be difficult to predict in detail, considerable variations (prediction errors) will remain even after a good prediction scheme has been applied. Furthermore, as the cell size has to be smaller to cope with the increased traffic, variations of traffic and interference among the cells becomes large. To overcome the problems caused by the fluctuations, one may rely on real-time measurements of the traffic situation, the propagation, and interference. The resulting real-time resource allocation is usually referred to as adaptive or dynamic channel allocation (DCA). With DCA, all channels are placed in a common pool and dynamically assigned according to the traffic and the interference situations. Hence, DCA is able to support traffic hot spots, borrowing the unused capacity from the neighboring cells. From a teletraffic point of view, FCA scheme behaves like a number of small groups of servers, whereas DCA provides a larger server. Thus higher utilization of channels can be expected from DCA for a given call blocking probability (trunking gains).

Contrary to conventional handover (more precisely termed as intercell handover), intracell handover refers to switching the channel of an ongoing user to the new one without change of base station (channel reassignment). In early investigation on DCA, intracell handover was not taken into consideration. However, by the introduction of the intracell handover, performance improvements have become possible in three aspects: increased adaptability to traffic, increased channel reuse and adaptability to interference. Figure 7.1 illustrates an attempt made by Beck and Panzer to roughly classify different schemes for DCA [1]. The classification is by no means rigorous but still serves the purpose of conveying the main principles of DCA and their interrelations. In the diagram, channel allocation schemes are represented by points in space. The origin in this figure represents conventional static (FCA)

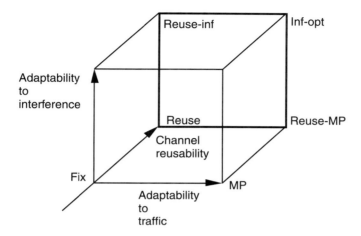

Figure 7.1 Classification of Dynamic Channel Allocation algorithms according to Beck and Panzer [1].

channel allocation. An algorithm far away from the origin has a high degree of adaptability, an algorithm close to the origin has a limited capability of coping with real-time fluctuations. The direction of the point from the origin corresponds to the different types of knowledge a DCA algorithm may exploit in its decision process.

The channel allocation may be dependent on the number of active users in the different cells, which has been the initial motivation of DCA. Such schemes are usually referred to as traffic adaptive channel allocation schemes. This property is represented by the traffic axis in Figure 7.1. Going to the extreme right corner of the cube (the point MP) corresponds to an algorithm exploiting the knowledge of the number of active users in every cell in an optimum way. Using information about the received signal strength (power level) corresponds to an algorithm on the reuse axis. Measuring background noise and interference is typical for algorithms on the interference axis.

A real-life resource allocation scheme uses combinations of these measurements and may thus be placed at some point within the cube. The best possible resource allocation scheme, which is capable to use all available information in the best way, is denoted INF-OPT and is located at the far upper corner of the cube. Unfortunately such an algorithm is believed to be very complex. In fact, no algorithm is known that performs this optimum channel allocation substantially faster than searching all combinations of channel assignments. Since the number of possible allocations grows (in the worst case) exponentially with the size of the problem, a brute force approach is impractical. Another important restriction in practical realizations of DCA schemes is that the dispersion of (measurement) information is not free; it will require transmission capacity and the information will be subject to delays. In many cases one will be therefore interested in studying schemes that are based on the decision of locally obtained information, that is, distributed DCA schemes, as opposed to often hypothetical centralized DCA schemes that may use any information freely in all access ports and mobiles. The latter schemes may not be very interesting from a practical point of view, but they are useful in the analysis since they provide us with upper bounds on the network performance.

Simulation and analysis over the last decades show that under low and nonuniform traffic density, DCA schemes give lower call blocking probabilities than FCA, whereas FCA performs better in high and uniform traffic density. The arrivals of new call requests incoming from cell to cell are in a random fashion. Therefore when DCA is used, different channels are assigned to serve calls at random too. Because of this randomness, the average reuse distance in DCA is likely to be greater than that in FCA. On the other hand, in FCA a specific channel can be assigned based on the minimum reuse distance D. Therefore, as the traffic density increases, the so-called trunking gain in DCA will be dominated by the randomness in channel reuse and FCA will become superior to DCA (see Figure 7.2)

Another comparison between DCA and FCA can be made in terms of forced call termination rate. As in Figure 7.3, with FCA, a user must be

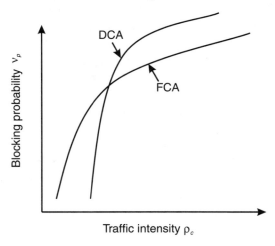

Figure 7.2 Call blocking probabilities of FCA and DCA as a function of call intensity.

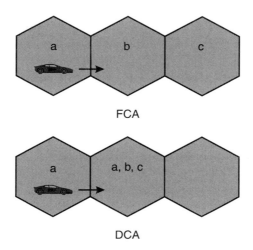

Figure 7.3 Call dropping in DCA.

handed over to another channel at every handover instance (from channel a to b in the figure) because the same channel is not available in adjacent cells. The channel assignment for the handover call is made in the same way as for a new call. Therefore, the call dropping (forced call termination) rate is similar to the blocking rate. In DCA, the same channel (channel a) can be used for the handover call unless this makes any interference problem, which may result in a lower forced call termination rate.

An important drawback of DCA schemes are, since no channels are assigned beforehand, every access port has to be physically capable of transmitting and establishing links on all of the C channels of the system. This is in contrast to the static allocation where the access port hardware may be confined to operate only on the assigned $\eta \approx C/K$ channels. In FDMA systems this may be a considerable drawback since normally each channel is associated with one specific piece of hardware (channel unit). In order to reach a compromise between capacity gains and hardware cost, a common method is essentially to revert to fixed allocation but in addition to the static channel set, maintain a small channel pool in which the channels may be preassigned to some access port but may be borrowed from a neighboring access port in case of threatening assignment failures (channel borrowing).

In the following, some generic DCA schemes will be investigated in order to gain some understanding how the different types of measurement information may be used. As has been seen in previous chapters, the general resource allocation problem involves the selection of waveform (channel), choosing the right access port and selecting a suitable power. Good resource allocation schemes will make all these choices simultaneously. To get a better understanding of the mechanisms involved in DCA, it will however be assumed that a (fixed) power level has been chosen and an access port has been assigned to every active user, leaving the problem of selecting the right channel (i.e., the channel allocation).

7.2 Traffic Adaptive Channel Allocation

In traffic adaptive channel allocation, one tries to adapt the allocation of spectral resources among cells in accordance with the current (measured) number of active mobiles in each cell. A (purely) traffic adaptive DCA scheme uses the kind of interference analysis that is the same as in the conventional static cell planning (FCA). Instead of splitting the available channels into channel groups of fixed size and statically assigning them to each access port, the dynamic scheme makes no advance allocations at all. Using a worst-case design, propagation conditions may be very roughly described using the cell compatibility concept. For any pair of cells, the worst-case design either ensures interference free operation or not. The compatibility properties may be summarized in the compatibility matrix, $\mathbf{I}$, with elements defined as

$$I_{ij} = \left(\begin{array}{l} 1, \text{ if active link in cell } i \text{ interferes with a link in cell } j \\ 0, \text{ otherwise} \end{array}\right)$$

Two cells, i and j, that are not compatible ($I_{ij} = 1$) may not use the same channel. The worst-case design ensures interference free channel allocation as long as the compatibility criterion is not violated. However, as we will see in the following, the main problem is that the number of active mobiles may be larger than the number of channels that can be assigned under these constraints. Note that the flexibility of the pure traffic adaptive system is substantial—for instance, all channels may be assigned to the same access port should this be necessary.

The optimum traffic adaptive channel allocation scheme, that is, the algorithm that minimizes the assignment failure rate Z^*, is the so-called maximum packing (MP) scheme. This policy assumes that a new call will be blocked only if there is no possible channel allocation (including intracell handover) to calls that would result in room for the new call. In other words, MP is a strategy to find the minimum number of channels to carry instantaneous existing calls while satisfying the compatibility constraints. If the required number of channels by MP is larger than the total number of channel C, then the new call will be blocked.

Defining a node as a cell (RAP) and an edge as a compatibility constraint, a cellular network can be represented by a graph (see Figure 7.4). If two cells, i and j, are not compatible ($I_{ij} = 1$), then there is an edge connecting nodes i and j. For the cell that has multiple users, we simply duplicate the node as many as the number of the users in that cell, and connect the duplicated cells by edges. With this graphical representation, MP can be interpreted as minimizing the number of colors that fill every node in a way that two adjacent connected nodes cannot be colored with the same color. This problem is called graph coloring problem, which unfortunately belongs to the class of NP-complete problems. Therefore, it is rather hopeless to find a fast (polynomial time) algorithm for finding the MP solution in general case.

Even with the vast reduction of complexity obtained by introducing the compatibility matrix, finding an optimal allocation, the MP solution, is not trivial. This is difficult even though, in the general case, there are many such allocations yielding the same number of assignment failures Z^*. Except in some interesting special cases [2], no solution method has been yet identified that is substantially better than checking the compatibility con-

$$I = \begin{pmatrix} 1 & 1 & 1 & 0 & 0 \\ 1 & 1 & 1 & 0 & 0 \\ 1 & 1 & 1 & 1 & 0 \\ 0 & 0 & 1 & 1 & 1 \\ 0 & 0 & 0 & 1 & 1 \end{pmatrix}$$

(a)

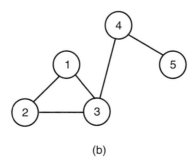

(b)

Figure 7.4 Graphical representation of compatibility matrix.

straints and evaluating Z for all possible channel allocations. The number of possible channel allocations is of the order C^B, that is, exponentially growing with the number of cells, B. Therefore, in practice, the MP solution provides only the performance bound that traffic-adaptive DCA can achieve, which is illustrated by the following example.

Example 7.1

A small cellular phone system consists of six access ports arranged according to the cell pattern in Figure 7.5. Assume that adjacent cells are incompatible, whereas other cell pairs may be considered to be compatible. The average traffic load is the same in all cells. Due to the random traffic fluctuations, however, assume that at some given instant the numbers of active mobiles in the different cells are

$$\mathbf{M} = \{M_i\} = \{7, 6, 2, 4, 8, 6\}$$

where M_i denotes the number of active mobiles in cell i. If the system has a maximum of $C = 18$ channels available, what will be the number of mobiles not assigned channel, Z, if

  a) A static cell plan is used?
  b) An optimal channel allocation is applied (maximum packing)?

Solution:

The compatibility matrix is easily derived from the conditions in the problem to yield:

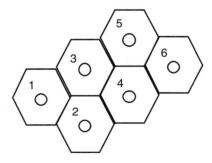

Figure 7.5 Wireless network example.

$$I = \begin{pmatrix} 1 & 1 & 1 & 0 & 0 & 0 \\ 1 & 1 & 1 & 1 & 0 & 0 \\ 1 & 1 & 1 & 1 & 1 & 0 \\ 0 & 1 & 1 & 1 & 1 & 1 \\ 0 & 0 & 1 & 1 & 1 & 1 \\ 0 & 0 & 0 & 1 & 1 & 1 \end{pmatrix}$$

a) It is rather easy to convince oneself that a symmetrical hexagonal cell pattern with $K = 3$ yields the lowest number of channels (cf previous chapters). The number of channels per cell becomes thus

$$\eta = C/K = 18/3 = 6 \text{ channels/cell}$$

The assignment failure vector $Z = \{Z_i\}$, where Z_i denotes the number of assignment failures in cell i, becomes

$$Z = \{1, 0, 0, 0, 2, 0\}$$

and

$$Z = \sum_{i=1}^{6} Z_i = 3$$

b) One may show that the following assignment, $A = \{A_i\}$, where A_i denotes the number of channels assigned to cell i,

$$A^* = \{7, 6, 2, 4, 8, 6\}$$

is feasible, that is, does not violate the compatibility constraints. For this case,

$$Z^* = \sum_{i=1}^{6} Z_i = 0$$

Here, the cells are numbered in such a way that the compatibility matrix exhibits a band structure. For example, this is the case if the cells are all placed along a line (highway system). If the ones in the compatibility matrix form a compact band on the diagonal of the matrix, there exists a simple algorithm to find the MP solution. This algorithm, the Gready algorithm, finds an optimal solution in linear time in B steps. To define the algorithm, we will use the auxiliary vector $Y = \{Y_i\}$, where Y_i denotes the number of channels that are still available for assignment in cell i.

The Gready Algorithm [2]

1) $A_1 = \min(M_1,\ C)$

● ● ● ● ● ●

i) $Y_i = C - \displaystyle\sum_{j=1}^{i-1} I_{ij}A_j \qquad 1 < i \leq B$

$A_i = \min(M_i,\ Y_i)$

Exercise:

Check that the algorithm above yields the same result as in Example 7.1(b).

Since the number of active mobiles is varying, the ideal (hypothetical) traffic adaptive DCA will reassign channels to each user according to the MP solution to each traffic variation. However, because of the large amount of measurement effort and high computational complexity for getting the MP solution, in reality, heuristic (suboptimal) algorithms are used as a compromise. Traffic adaptive channel allocation schemes (heuristic algorithms) are very effective when the traffic fluctuations are very large relative to the number of channels available. In particular, this situation is encountered when the traffic load is low to moderate and the average number of channels per cell η can be kept low (<10–20). Capacity gains (compared to static allocation) in the order of 30–50% are not unusual. For systems with heavy traffic and a large number of channels, however, the relative improvement is reduced, due to the fact that the relative variation in traffic decreases and the static cell plan will provide channel allocation that is close to the optimum. DCA assigns channels to cells in advance, so that the average reuse distance is likely to be greater than that in the static channel allocation.

7.3 Reuse Partitioning

An important drawback of pure traffic-adaptive schemes is that they rely on the worst-case analysis that is the basis for determining the compatibility matrix. To ensure compatibility (interference-free communication), it must be possible to guarantee a sufficient SIR under all conditions (i.e., mobile positions, propagation phenomena, interference/load conditions, and so forth). Such a design will require large margins regarding the reuse distance and will likely be rather inefficient. Here we attempt to remedy this problem by introducing DCA schemes adapting to the propagation conditions. Again, the static channel allocation is used as the starting point. The choice of the number of channel groups is done either by means of the worst-case analysis, satisfying the SIR-requirement on the perimeter of a cell, or by means of the outage probability criteria described in the previous chapters. As previously noted, the average SIR drops as the cell boundary is approached. Maintaining a sufficiently low average outage probability in the cell, mobile terminals close to the access port in general reach a much higher SIR than the required γ_0, simply due to the fact that the received signal level is much higher for these terminals. Such a terminal is thus able to tolerate a much higher interference level and still satisfy the SIR requirement γ_0. Rephrasing this statement would be to say that terminals in the interior of the cell would tolerate a lower reuse distance. This is the basic idea behind the Reuse Partitioning (RP) strategy [3]. Instead of using a single cell plan, several overlaid cell plans with different reuse distances are combined.

Terminals close to the cell center (high signal level) choose channels from a channel set in a cell plan with short reuse distance, whereas terminals at the cell perimeter (low signal level) choose channels from a cell plan with larger reuse distance. Let us assume that a total of L different cell plans are

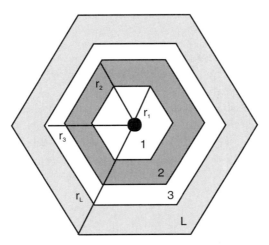

Figure 7.6 Zones in reuse partitioning.

used with cluster sizes K_1, K_2, ..., K_L. Without loss of generality, we number the cell plans such that

$$K_1 < K_2 < \ldots K_L$$

Choosing the cell plans from the class of symmetric hexagonal cell plans, the most general case is that $K_1 = 1$ and $K_L = K$. A cell plan with K_i channel groups will have a reuse distance D_i given by

$$D_i = R\sqrt{3K_i}$$

In order to study the basic principle of the allocation scheme, neglect the fading phenomena momentarily and assume that the received signal power is only distance dependent (inverse power law). Using this assumption, the signal power may be computed at a terminal at distance r from the access port at high load

$$\Gamma^u(r) \approx \frac{D^4}{7r^4} = \frac{9R^4}{7r^4} K_i^2$$

The condition $\Gamma^u(r) \geq \gamma_0$ now yields

$$r < R\left(\frac{9}{7\gamma_0}\right)^{1/4} \sqrt{K_i} = r_i$$

A cell plan with cluster size K_i can thus be used by all terminals up to a distance r_i from the cell center (the access port). A terminal in zone i may use a cell plan with cluster size K_i (or larger).

Having established the SIR in various parts of the cell, now turn to the actual channel allocation. In a pure RP scheme this is done by simply determining the number of channels that is to be allocated to each cell plan. For this purpose, define the capacity allocation vector as

$$\boldsymbol{c} = (c_1, c_2, \ldots, c_L)$$

where the ith component c_i is the number of channels assigned to each channel group in cell plan i. The choice of capacity allocation vector is constrained by the total number of available channels C by the relation

$$\sum_{i=1}^{L} c_i K_i \le C$$

The total number of channels that will be available in a cell is

$$c_0 = \sum_{i=1}^{L} c_i$$

Note that only the terminals in the center of the cell will have access to all these c_0 channels, whereas terminals on the perimeter are limited to use one of the c_L channels of the outermost zone. Finding the optimal $\mathbf{c}$ (for lowest assignment failure probability) is a rather straightforward but tedious optimization problem.

The RP procedure is effectively splitting the existing cell pattern into smaller areas. The average number of channels available to the mobiles will increase, but due to the fact that there will be a smaller number of terminals in each of the zones, the relative traffic variations will actually increase. Numerical results show that for high traffic loads the former phenomenon will dominate and lower assignment failure rates than the conventional static scheme (allocating all channels to the outermost zone, $c_L = \eta$) are obtained. For low traffic loads, however, the traffic variation will dominate and worse performance than the static scheme will be achieved (trunking loss). To obtain a good performance out of the RP scheme, it is necessary to combine it with a traffic-adaptive scheme. Instead of confining a terminal in one zone to use the channels dedicated to a particular zone i, observe that this terminal is also able to use channels allocated to zones further out towards the perimeter, that is, $i + 1$, $i + 2$ ... L. If the number of active terminals exceeds the allocation to their zones, channels may be borrowed from zones further out. It can be shown that the following simple channel assignment algorithm minimizes the assignment failure rate [4]. Assume that a channel allocation $\mathbf{c}$ has been made. For any given traffic pattern $\mathbf{M'} = \{M_i\}$, where M_i denotes the number of active terminals in the zones $i = 1, 2 \ldots L$, assign channels for one zone at a time.

Optimal RP Channel Allocation [4]

1) Assign $\min(M_1, c_1)$ channels from cell plan 1 to terminals in zone 1. Let $Z_1 = \max(0, M_1 - c_1)$.

i) Assign $\min(M_i + Z_{i-1}, c_i)$ channels from cell plan i to terminals in zone i or zones with a lower index. Let $Z_i = \max(0, M_i + Z_{i-1} - c_i)$.

L) Assign $\min(M_L + Z_{L-1}, c_L)$ channels from cell plan L to terminals in zone L. Let $Z = Z_L = \max(0, M_L + Z_{L-1} - c_L)$.

The vector $\mathbf{Z} = (Z_1, Z_2, \ldots Z_L)$ describes how many channels have to be borrowed form zones with higher index. The demand for additional channels in the last (outermost) zone, Z_L, cannot be satisfied and thus becomes the total number of assignment failures, Z.

Note the similarity of the scheme to the Gready algorithm. Intuitively it can be understood that this procedure is optimal, since by using the Gready algorithm, it is ensured that a channel is never saved for later (for the outer zones) as long as there is an active mobile requiring it. Saving a channel in that situation will always expose the risk that it is never used later despite the fact that it is known that there is at least one terminal needing it. The performance of some RP schemes is illustrated in Figure 7.7, which shows the assignment failure rate as a function of the traffic load for a system with $C = 100$ channels and 5 zones corresponding to $K = (1, 3, 4, 7, 9)$. The lower curves show the performance of the optimum (Gready) channel assignment scheme for the capacity allocations: a) $\mathbf{c} = (2, 4, 2, 6, 4)$; b) $\mathbf{c} = (3, 3, 2, 5, 5)$; and c) $\mathbf{c} = (2, 2, 2, 3, 7)$. The top curve is the assignment failure rate for static assignment system with $K = 9$, the second curve from the top shows the performance of a system with the same capacity allocation as in a) but with no borrowing. The graph shows that at the 1% level of assignment failure rate, the RP schemes yield a capacity increase in the order of 50% compared to static allocation (no reuse). As a reference, the graph also shows the performance of a pure RP (no borrowing). Due to the trunking loss for low loads ϖ_c, this scheme yields a performance that is worse than the static scheme.

It can be seen that the optimum choice of $\mathbf{c}$ depends on the traffic load. Due to the integer constraints on the components in $\mathbf{c}$, the optimum $\mathbf{c}$ is found by an exhaustive search over a rather moderate number of feasible vectors. A decent choice of $\mathbf{c}$ for high loads and a moderate to large number of channels C are to choose the c_i close to proportional to the expected number of terminals in each zone. If the terminals are uniformly distributed, this is the same as choosing the c_i proportional to the zone areas, that is, accordingly

$$c_i = \frac{C(K_i - K_{i-1})}{\sum_{j=1}^{L} (K_j - K_{j-1})K_j}$$

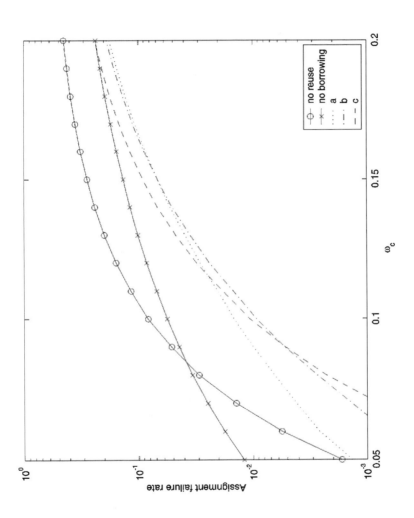

Figure 7.7 Assignment failure rate as function of the traffic load ϖ_c for a system with $C = 100$ channels and five zones corresponding to $K = (1, 3, 4, 7, 9)$.

Exercise:

Show that c_i given by the above equation is indeed proportional to the area of zone i. Which of the vectors c in Figure 7.7 is chosen this way?

For low loads and for low numbers of channels C, Figure 7.7 indicates that more channels should be allocated to the outer zones than given by the above equation. This creates a larger flexibility (channels far out may be used by almost all terminals) at the expense of a lower total number of channels c_0 in the cell. In an overload situation, an attempt should be made to maximize c_0 by allocating more channels to the interior zones (sacrificing the terminals at the perimeter). Figure 7.8 shows that the number of zones used does not have to be very large to get substantial performance gains. Already 25–30% of capacity improvement is obtained by going from a conventional static scheme to a two-zone RP-scheme.

For the sake of simplicity, the RP scheme has been analyzed with a simple distance dependent propagation model (e.g., not including shadow fading). However, this is not a serious restriction. In practical implementation, the algorithm will certainly use the received signal level to determine to which cell plan a terminal should be assigned. It is of no consequence that in the fading case there is no direct correspondence between geographical position and signal level.

7.4 Interference-Based DCA Schemes

Both the traffic-adaptive and the RP schemes are based on conservative (pessimistic) estimates of the propagation and interference situations. The compatibility matrix forming the foundation of the traffic-adaptive schemes, are based on pair-wise cell compatibility. This means that two cells compatible of two connections may be established on the same channel in these two cells, regardless what the interference situation caused by links in other cells may be. In principle this also holds for the RP schemes, which in their pure forms are nothing else than refined static allocation schemes. Not knowing the interference power forces us to use large signal level margins in order to guarantee an sufficient signal quality resulting in capacity losses. One way to resolve this problem would be to use a channel assignment that is based on the actual interference level, rather than some statistical estimates. Some simple and straightforward algorithms to exploit this concept not only have been proposed in the literature, but are also used in real-life cellular communication systems (e.g., the DECT cordless PABX system). The algorithms

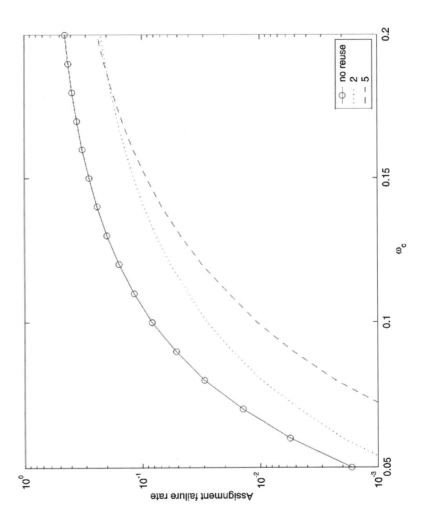

Figure 7.8 Assignment failure rate as function of the traffic load ϖ_c for a system with $C = 100$ channels and $L = 2$ and 5 zones, respectively.

proposed are usually distributed, that is, terminals and access ports themselves select channels based on their own observations of the interference situation. The simple, distributed character makes these schemes attractive from an implementation point of view. In the following, a few examples of heuristic algorithms based on the estimation of the local SIR will be shown.

A simple, but effective, scheme is proposed by Beck and Panzer [1]. The main concept is illustrated by the flow chart in Figure 7.9. The access ports continuously monitor or estimate their SIRs and compare this SIR estimate with a threshold γ_c. The measurement may also be made by the terminal (downlink) or by both at the same time (e.g., compare the minimum SIR with the threshold). If the SIR estimate drops below the threshold γ_c (i.e., the signal quality becomes too low), the access port will start scanning for a new channel with better interference conditions. The channels are searched sequentially until a new channel with a projected SIR is larger than γ_k is found. The SIR projection is done by measuring the current interference and comparing this with the current received signal level. If such a channel is found, the connection is moved (intracell handover) to this channel; otherwise, the old connection is kept intact.

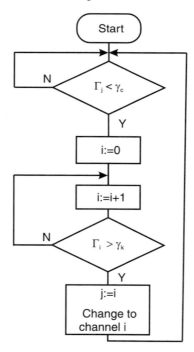

Figure 7.9 Simple distributed dynamic channel allocation algorithm based on estimated SIR. The terminal is initially assumed to be using channel j.

The choice of the thresholds γ_c and γ_k determines the performance and the stability of the algorithm. The threshold γ_c is normally placed somewhat above γ_0 and provides a certain time margin to perform the channel swap if the SIR is dropping rapidly. A high value of γ_c provides an ample margin, but the algorithm becomes nervous with frequent changes of channel. If the channel allocation procedure first compared to the fading phenomena is encountered, the threshold γ_c can be placed very close to γ_0. For the threshold γ_k, in a sense, the converse is true. A low threshold γ_k results in easily finding a new channel that fulfills the SIR requirement. On the hand, not being choosy yields a quite low SIR in the new channel. Changing to a channel with low SNR usually means that there are already many other terminals using this channel. If our link starts by using the channel, there is a risk that some of the other terminals are dropping below γ_c, forcing them to change the channel. These changes may in turn result in new changes. A high threshold γ_k reduces the risk of an avalanche of channel changes and thus promotes the stability of the system. Algorithms with a high γ_c and a low γ_k aggressively search to improve the signal quality resulting in a system with sudden SIR changes and frequent channel swapping. Conversely algorithms with a low γ_c and a high γ_k represent conservative designs where one tries to avoid unnecessary channel swapping by postponing channel changes as long as possible. If forced to change channels, the algorithm selects safe channels in which the probability of causing interference to surrounding terminals is low.

Another similar but slower and conservative method is the channel segregation scheme proposed by Furuya and Akaiwa [6]. Here an access port uses interference measurement to update a priority vector $\mathbf{Q} = (Q_1, Q_2, \ldots, Q_C)$ where the integers Q_i describe the priority of channel i. When a channel is required, the channel with the highest priority estimates the would-be SIR. If this SIR estimate is below a threshold γ_k, the priority of this channel is reduced by one and the channel with the second-highest priority is investigated. If the SIR exceeds γ_k, the channel is used and the priority is increased by one. The result is that the access port "learns" which channels usually work well. Simulations show that the algorithm is usually very stable (i.e., the different access ports find their own channels and rarely compete for the same channels).

Problems

7.1 To achieve a high capacity in the downtown area of a big city, nine microcells have been installed, as shown in Figure 7.10. The modulation and coding schemes are so robust that the same channel may be reused

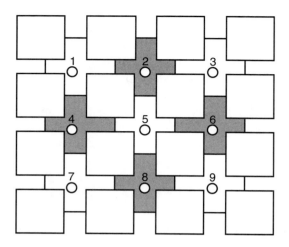

Figure 7.10 Cell structure in downtown area.

in every other cell along a street, independently of the mobile locations. Interference from cells that are not on the same street can be neglected completely due to the high diffraction loss. In the citywide cell plan C_{micro} = 10 channels have been allocated to these microcells. Currently, we have the following distribution of active terminals:

$$M = (2, 6, 2, 3, 4, 8, 0, 8, 1).$$

a) Determine the minimum number of terminals that are not assigned a channel if static channel allocation is employed.
b) Determine the number of terminals that are not assigned a channel if the Gready algorithm is used. Does it matter which actual channels a terminal is using?
c) Is there a better assignment than the one achieved by the Gready algorithm? If this is the case, determine the number of terminals that are not assigned a channel for this assignment.

7.2 To provide a high capacity along a section of a highly congested highway, 10 microcells of equal size have been installed. The modulation and coding schemes are so robust that the same channel may be reused in every other cell along the highway, independently of the locations of the terminals. In the citywide cell plan C_{micro} = 20 channels have been allocated to these microcells. Currently, the following traffic situation prevails:

$$M = (8, 11, 4, 14, 9, 15, 3, 13, 8, 15).$$

a) Determine the minimum number of terminals that are not assigned a channel if static channel allocation is employed.
b) Determine the number of terminals that are not assigned a channel if the channel borrowing scheme is employed. Is it of any importance which channel is assigned to a specific mobile?
c) Determine the number of terminals that are not assigned a channel if the Gready algorithm is used. Does it matter which actual channels a terminal is using?

7.3 A cellular communication system utilizes a reuse-partitioning scheme with two zones (see Figure 7.11). A total of 100 channels are available and minimum SIR of 16 dB is required on the perimeter of each zone. The cell radius is $R = R_2 = 1$ km. The (average) propagation path loss is assumed to be proportional to the fourth power of the distance.

a) Determine the cluster size (number of channel groups) that is required for the outer zone (zone 2).
b) Compute the *maximum* radius of the inner zone if we choose to use the cluster size 3 in the corresponding cell plan.
c) Compute the capacity allocation vector **c** that assigns channels in proportion to the area of each zone.

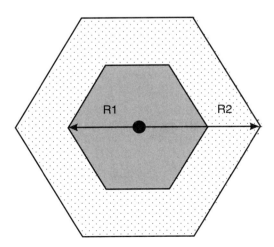

Figure 7.11 Cellular communication system with two cells.

d) Determine the primitive channel assignment failure rate as a function of the traffic intensity (without intracell borrowing).

e) Compute the corresponding primitive channel assignment failure rate for a conventional system without reuse-partitioning. For which traffic loads does the conventional system outperform the reuse-partitioning system?

7.4 Consider a small cellular wireless communication system with only three cells and only three (duplex) channels. In a certain moment there are only three active terminals in the system. The propagation situation can be described by the link gain matrix.

$$G = \begin{pmatrix} 2 \cdot 10^{-6} & 0.1 \cdot 10^{-6} & 0.1 \cdot 10^{-6} \\ 0.1 \cdot 10^{-6} & 1 \cdot 10^{-6} & 0.2 \cdot 10^{-6} \\ 0.8 \cdot 10^{-6} & 0.2 \cdot 10^{-6} & 3 \cdot 10^{-6} \end{pmatrix}$$

Assume that we choose our numbering scheme in such a way that terminal i is connected to access port i using channel (pair) i, $i = 1, 2, 3$. Now a fourth terminal becomes active with the following link gains to the access ports:

$$G_{i4} = \begin{pmatrix} 3 \cdot 10^{-6} \\ 0.2 \cdot 10^{-6} \\ 0.3 \cdot 10^{-6} \end{pmatrix}$$

a) Assume that a DCA scheme is used in which the new terminal will choose the channel which it perceives will provide the highest SIR. Which channel will the new terminal choose?

b) Which channel would be selected by a centralized SIR adaptive scheme maximizing the minimum SIR for all mobiles?

References

[1]   Beck, R., and H. Panzer, "Strategies for Handover and Dynamic Channel Allocation in Micro-Cellular Mobile Radio Systems," *Proc. Vehicular Technology Conference (VTC'89)*, San Francisco, 1989.

[2]   Frodigh, M., "Dynamic Channel Assignment in 1-Dimensional Cellular Radio Systems," *Licenciatavhandling*, TRITA-RCS-9201, Royal Institute of Technology, 1992.

[3]   Whitehead, J. F., "Cellular Spectrum Efficiency Via Reuse Planning," *Proc. Vehicular Technology Conference (VTC'85)*, 1985.

[4]   Zander, J., and M. Frodigh, "Capacity Allocation and Channel Assigment in Cellular Radio Systems Using Reuse-Partitioning," *Electronics Letters*, Vol. 28, No. 5, February 1992, pp. 438–439.

[5]   Goodman, D. J., S. A. Grandhi, and R. Vijayan, "Distributed Dynamic Channel Assignment Schemes," *WINLAB Technical Report*, New Jersey: Rutgers University, August 1992.

[6]   Furuya, Y., and Y. Akaiwa, "Channel Segregation—A Distributed Adaptive Channel Allocation Scheme for Mobile Communication Systems," *Proc. DMR II*, Stockholm, 1987.

8

Orthogonal Frequency Hopping

8.1 Random Channel Allocation

Frequency hopping (FH) was introduced in Chapter 2 primarily as a means to counter multipath fading. The most straightforward application is orthogonal frequency hopping, where synchronous, nonoverlapping hopping patterns are used. The performance of this technique was found to be equivalent to any other orthogonal multiple access schemes in the multi-access channel model introduced in Section 2.1. The latter model could be used to describe the up-link in a single cell in a wireless network. In this chapter, orthogonal frequency hopping will be investigated more closely in a more general multiple cell environment when hopping pattern orthogonality may be kept within the group of terminals accessing one access point, but not with terminals using other access points. As was demonstrated in Chapter 2, frequency hopping combined with error control coding provided very effective protection against interference. This will prove to be the case also in more general wireless network settings. The approach taken in this chapter will be to demonstrate that orthogonal frequency hopping can be interpreted as a conventional channel allocation scheme, where the channel allocations are rapidly permuted. All statistical results regarding outage probabilities from Chapter 4, will therefore be directly useful in the performance analysis of orthogonal FH systems. Finally, it is demonstrated that the single most important element of proper FH system design is the error control scheme. Some effort is therefore devoted to reviewing some of the key issues in this field.

In the section on static channel allocation in fading environments, (see Chapter 4), a strong connection between the assignment failure rate ν and the interference rate χ was found. By choosing a small number of channel groups K, each cell will be assigned a large number of channels and the event where a cell runs out of channels will have a low probability. On the other hand, the interference level will increase, and frequently an assigned channel will turn out to be useless due to excessive interference encountered on it. This is illustrated in Figure 4.14 for a system with $C = 100$ channels. In a mobile telephone system, a user is not likely to tolerate being deprived of a channel or experiencing poor transmission quality. This means that the probabilities of both these events have to be very small in a conventional static assignment system. As can been seen from the design examples in Chapter 4 and in real mobile telephone systems, these requirements lead to designs with fairly large cluster sizes K. The latter parameter ranges from 19 or 21 in analog systems to 9–12 in the digital systems. The performance of the resulting systems is mainly limited by the assignment failure rate ν, which rises much more steeply than the interference rate χ (blocking limited systems). The result is poor capacity figures.

In order to get substantially higher capacity, it would be required to somehow improve the capabilities of the system to withstand interference. Using better modulation and coding schemes in order to push down the required SIR γ_0 has been successfully used in the transition from analog to digital systems. With the current level of sophistication in coding and signal-processing schemes already operating at performance levels close to the theoretical bounds, there is not much hope for yet another quantum leap in performance improvement along this general path. Instead some very specific knowledge about the interference environment should be exploited. Study, for instance, Figure 8.1 which shows the typical behavior of the SIR:s in a set of five terminals in a mobile wireless system with static channel allocation. The upper graph shows the SIR for slowly moving terminals whereas the lower graph illustrates the situation for fast moving terminals. The outage rate, that is, the time the terminals collectively spend below the SIR threshold γ_0, is the same in both examples, but the behavior of the SIR is quite different in the two cases. In the slow moving terminal case, one of the terminals remains below the threshold all the time, whereas the others all have favorable conditions. In most cases, the outage rate will be the fraction of terminals that are unable to communicate. In the fast moving terminal case, all terminals will take their share of the outage problem. Here the outage rate in fact describes the fraction of time a terminal is unable to communicate. If the terminals move fast enough the outage periods may be

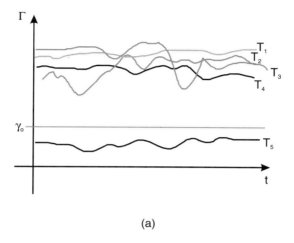

(a)

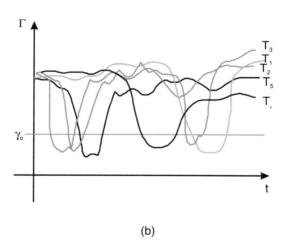

(b)

Figure 8.1 Typical realizations of terminal SIR:s in cellular systems with slowly moving terminals (a) and with rapidly moving terminals (b).

short enough to be handled by error control coding. In this case there could be an allowance for some coding redundancy in order to combat the interference.

Since the system designer cannot control the mobility of the terminals, other ways to emulate this behavior have to be found. One method that mimics mobility even for stationary terminals is to frequently change channels in a random fashion. We call this procedure *random channel allocation*.

Changing channels will make the terminals (and access ports) to meet a different (independent) interference situation in each hop. It is important to note that the outage probability χ is not affected at all by this operation. Remember that this quantity is the probability that a randomly picked terminal, at some randomly selected instant, experiences a link SIR less than the threshold. The permutation process will in each step simply create a new realization of the SIR levels in the different links, but not change their probability distributions. This means that the permutation process in itself does not improve the performance of the system but the resulting SIR realizations, the outage process, will become easier to handle.

If the channels are in fact implemented as narrowband (FDMA) signals, this exactly corresponds to a frequency hopping system (see Chapter 2). This is no doubt the most common practical realization of the random allocation concept. However, it is obvious that any other permutation of sets of orthogonal waveforms will have the same property, for example, time hopping (TH) = permutation of time slots in TDMA systems. In the next chapter, it will be demonstrated that also the nonorthogonal CDMA waveforms can be used in a similar fashion. FH systems, however, have the additional benefit, which is not shared for instance by TH schemes, in that they exhibit good performance in frequency selective fading.

Frequency hopping can be based on a system with static channel allocation. Instead of assigning the terminals, a single channel in the channel set in the channel group assigned to the corresponding access port, we rapidly permute the channel assignment over the entire channel set in the channel group assigned to that access port. (see Figure 2.11). The time interval where the terminals and access ports remain at the same frequency was in Chapter 2, coined a *chip*. Each terminal will be assigned a new channel out of the set in each chip. Doing the permutations (hopping) synchronously for all terminals in a cell maintains orthogonality (interference-free communication) within the cell. Also, the hopping procedure will not cause any new interaction with the nearest neighboring cells since these are protected by the frequency reuse constraints and use other channel sets, exactly in the same way as in the underlying static channel allocation system. The hop rate of such a system cannot be very high due to the difficulty of maintaining synchronism between the terminals and the access port. The orthogonal hopping technique is thus confined to use slow frequency hopping (SFH) where whole messages (blocks, bursts) are transmitted in each chip.[1]

1. Faster hopping and the use of non-orthogonal hopping sequences (allowing more users) have been discussed frequently in the literature [12] but have so far found little practical application in civilian systems.

Since the design of the error-control scheme is critical to the performance of an FH-system, a short review of some basic facts in coding theory is appropriate at this time. The reader unfamiliar with these concepts is referred to [1] or [2] for a more thorough analysis. In an SFH-system, a block that is transmitted on a channel may experience excessive interference and may be lost—not just a few symbols, but rather a complete block of symbols completely lost. Any successful error control scheme has to utilize this characteristic. Figure 8.2 illustrates the stream of blocks as received by, say one terminal, in one such a scheme. The data is formatted and encoded into blocks of size n_B symbols. Each of these blocks has thus been transmitted on a different channel and has been subject to different interference conditions. L such blocks form a code word, that is, the total code word length is $n_B L$ symbols. If the outage probability is χ, this means that, on the average, χL of the received blocks in a code word will experience bad SIR and exhibit very high symbol error rates. A conservative estimate of the number of symbol errors in such a block will be that all symbols are in error, that is, lost. Any error control scheme with ambitions to be successful thus needs to be capable of correcting at least $\lceil \chi L \rceil$ blocks (or $t = \lceil \chi n_B L \rceil$ symbols) and preferably more due to the statistical variations of the error pattern. Also, in a practical scheme, consideration has to be given to the fact that, occasionally, there will be symbol errors also in those blocks received above the SIR threshold.

The potential performance of such a coding scheme is briefly investigated. In order to correct a certain number of erroneous symbols, redundant symbols have to be added in the encoding process. If in a code word of n symbols, k symbols are information symbols, $n-k$ symbols will be redundant or check symbols. The *rate* of such a code is

$$R = \frac{k}{n}$$

Now, from the basic theory of (block) codes, it is known that the number of errors a code can correct is related to its minimum distance (i.e.,

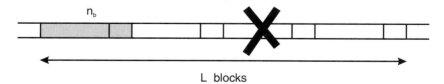

Figure 8.2 Block structure of error control scheme.

the minimum number of symbol positions in which two code words differ) d_{min} in the following way

$$t = \left\lfloor \frac{d_{min} - 1}{2} \right\rfloor \approx \frac{d_{min}}{2}$$

where the last approximation holds for large n and d_{min}. The questions arise, what maximum d_{min} can be achieved and what is the price paid in terms of added redundancy? A well-known performance bound is the Singleton bound that states

$$d_{min} \leq n - k + 1$$

Dividing by n yields

$$\delta = \frac{d_{min}}{n} \leq 1 - R + \frac{1}{n} \approx 1 - R$$

where again the last approximation holds for large n. Combining this with the error correction capability equation above yields

$$t/n \approx \frac{d_{min}}{2n} \leq \frac{1 - R}{2}$$

If it would be possible to design a code achieving the Singleton bound the relative error correction capability of this code would be

$$t/n \approx \frac{1 - R}{2} > \chi \Rightarrow R < 1 - 2\chi \qquad (8.1)$$

Codes of this type exist and are usually referred to as *Maximum Distance Separable* or MDS codes. The popular class of Reed-Solomon (RS) codes is of this type.

Now study (8.1) a little more carefully. Using a redundancy level that is roughly only twice the outage probability, it is possible to correct all the errors and thus communicate virtually error free. For a 10% outage probability, 20% redundancy would be required, leaving 80% of the link capacity available for payload data. This fact provides a revolutionary new way to treat the concept of outage. In fact things look even better, considering the fact that errors are confined to blocks. By using a (very small) part of the redundancy in each block for error detection, a very reliable indicator of which blocks are in error can be designed. Using this information, an erasure decoder (from classical coding theory) can be employed. Here unreliable/erroneous blocks are declared to be erased and the original message is reconstructed only from the reliable information. A code can be shown to correct

$$s = d_{min} - 1 \approx d_{min}$$

erasures. Using such an ideal erasure decoder would yield

$$s/n \approx 1 - R > \chi \Rightarrow R < 1 - \chi \tag{8.2}$$

This is investigated in more detail by considering the following example.

Example 8.1 Slow Frequency Hopping and Ideal Coding

A mobile communication system has access to bandwidth corresponding to 100 channels when coding is not employed. A block-synchronous slow frequency-hopping scheme is to be used. The frequency assignment avoids collisions within the cell and is random and independent of the frequency assignments in all adjacent cells. The propagation is assumed to be distance dependent according to the examples in Chapter 4 (propagation constant $\alpha = 4$, log-normal fading $\sigma = 8$ dB, noise negligible). The data is coded using an MDS code with rate $R = 2/3$. A block is correctly detected if the instantaneous received SIR during the block transmission is above 9 dB. The code is much longer that the chip duration and capable of perfectly detecting all block containing erroneous blocks by using a negligible amount of check symbols. We study the uplink (terminal-to-access point) of the system only and assume that perfect constant-received power (CRP) power control is exercised. Estimate the capacity of the uplink of this system and compare the result with a conventional static system where an assignment failure rate of 5% is required!

Solution:

Computing the C/I in some access point for some mobile terminal i connected to access port j

$$\Gamma_i = \frac{P_0}{\sum_k X_k P_k G_{kj} + N_0} \tag{8.3}$$

where $\mathbf{X} = \{X_k\}$ is the binary activity vector, and P_0 is the received power target for the CRP power control scheme. Notable is that the C/I is independent of i and thus the same for all terminals connected to access port j. Using a RUNE simulation (see Appendices B and C) provides the outage rate χ as function of the relative traffic load q on a channel as shown in Figure 8.3. According to (8.2) the error control scheme will work as long as

$$\chi < 1 - R = 33\%$$

In addition, the coded scheme will only have $C \cdot R = 67$ channels available due to the added redundancy. Computing the total assignment failure rate for different cluster sizes for the coded and uncoded system yields Table 8.1.

Note that the best coded system has virtually no assignment failures (it is almost completely *interference limited*) since the number of channels in each access port is so large and the relative traffic load is moderate. The capacity of the best SFH system is almost twice that of the conventional system.

As can be seen in the example, the SFH system designs are characterized by a large number of channels per access port but relatively low channel utilization. The latter is in some parts of the literature denoted fractional loading. The result is a system that is completely interference limited, and blocking or channel shortage in any cell is not an issue at all. In a sense there exists a fair system providing the same performance in the uplink to all users in the cell. The CRP algorithm is in fact providing a kind of "C/I-balancing" in this case. In the downlink the situation is different. (Theoretically a C/I-balancing scheme could be used here as well, which on the other hand, would not be a CRP scheme since the terminals all experience different interference situations). Due to the pseudorandom frequency hopping, it is virtually impossible to accurately estimate or predict the interference in the next time/frequency slot—and even more difficult to track the interference power with any reasonable precision. Instead, power control schemes using average interference measurements have been proposed [3, 4, 5].

Another important property of such an SFH random channel allocation scheme, is that it fragments the user data into blocks that are in general

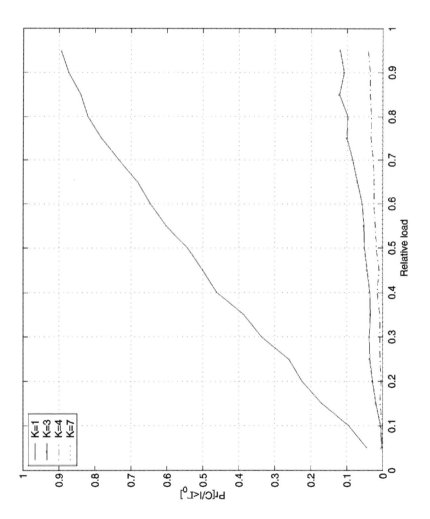

Figure 8.3 Uplink Interference/Outage probability as function of relative load factor q. $K = 1, 3, 7, 9, 12$ constant received power control (CRP).

Table 8.1
Results in Example 8.1

SFH System with Ideal Coding ($C' = 100R = 75$ Channels)					
K	η	q	ν_0	χ	$\varpi_c = \eta q$
1	67	0.13	0	25%	8
3	*22*	*0.55*	*<0.1%*	*25%*	*12*
7	9	0.7	5%	<25%	6
Conventional System No Coding ($C = 100$ Channels)					
K	η	q	ν_0	χ	$\varpi_c = \eta q$
9	*11*	*0.5*	*1%*	*4%*	*5.5*
12	8	0.4	2.5%	2.5%	3.2

much smaller than the data units of the payload traffic. Thus the interference from a speech user or from a data user will have very similar characteristics. Further, in a system with a low K (e.g., $K = 1$), the sum in the denominator of (8.3) consists of many terms. The (random) sum will become very close to its average and the variance is decreasing with the number of terms. The consequence of this is that it will be the individual users' average interference contribution that will be of importance, not its instantaneous value. For example, speech users using a discontinous transmission (DTX) mode of operation, will physically transmit only when actually speaking and keep quiet (no power emitted) otherwise. Since measurements have shown that a typical user is actually talking (and thus producing speech data) only 40% of the time (the so-called *voice activity* factor), the average interference power of such speech terminals is only 40% of that of a continuous transmitter. In theory, the DTX technique would thus increase the capacity by a factor of 2.5 compared to a conventional voice system. Similar gains can be obtained also for interactive class data traffic, where the peak to average data rate ratio is even larger. Some care has to be exercised in making these approximations. The limit theorem only holds when the number of currently active users is large. When increasing the peak to average ratios in the input data traffic stream, very few interferers may instantaneously dominate the interference sum, which in turn may exceed the average interference level by a considerable amount. A similar situation where this may appear is when highly directional antennas are used in the base station (see Section 4.5). Again, in this case the number of interferers is reduced, which decreases the average of the interference sum but also increases the relative variance of the sum.

8.2 Slow Frequency Hopping System Design

Obviously, there are some practical problems in achieving the promising performance demonstrated in Example 8.1. The results shown there hold for large code word sizes and since errors occur in blocks, this would mean that not only $n = n_B L$ but also L, the number of blocks has to be large. There are two main disadvantages involved: the increased complexity of encoders and, in particular, decoders, as well as the increased delay. From the literature it is known that designing codes that are good at correcting errors that are confined to certain blocks (error burst) is a much simpler task than correcting errors that are randomly dispersed over the code word [6]. The result is coding schemes with good performance achieved at rather moderate complexity. *Interleaving* is one popular technique where short, low complexity, codes can be combined to yield long codes with good block-error capabilities. *Concatenating codes* is another, more efficient, technique to the same effect. It is beyond the scope of this book to explore the details of code design and the reader is referred to [6] and [2] for these details.

The delay, however, is a more severe problem, in particular in speech communication systems. The decoder cannot make a decision until all L blocks in the code word have been received. Assuming that the hop frequency is f_h, we note that a delay of L/f_h seconds is inevitable. There will clearly be a trade-off between the error-control performance, the residual error probability (after decoding) which decreases with L, and the delay which increases linearly with L. An interesting observation is that coding also will become effective to combat momentary overload situations. If we have more active terminals than channels, we can let the terminals with some appropriate probability hop to an imaginary silent channel, that is, not transmit at all. The receiver will handle these missing chips in the same way as if they were lost due to fading or strong interference.

A practical example of a system that exploits orthogonal frequency hopping of this kind is the GSM system [7]. In GSM, a frequency hop takes place for every new burst, that is, every 4.615 ms, resulting in a hopping rate of 217 hops per second [8]. Hopping sequences of length 64 are specified. If the system has Q frequencies available, a total of $64Q$ sequences can be used in the same cell. The hopping sequences are described by two parameters, the hopping sequence number (HSN) and the mobile allocation index offset (MAIO). The HSN can take 64 different values, whereas the MAIO can take Q values. Two GSM channels using the same HSN but different MAIO will never use the same frequency in any time slot. In the opposite case, the hopping sequences are designed as pseudorandom sequences to allow two channels using different HSNs only to coincide in a single time slots in each sequence. One sequence, HSN = 0, corresponds to a cyclic hopping mode, where all terminals and base stations follow the same hopping pattern and will thus experience different multipath fading but the same interference on the different channels. This solution is to be avoided because the interference averaging or interference diversity effect discussed above is not utilized at all. Instead, all channels in one cell are using the same HSN and different MAIOs, again avoiding all interference inside the cell. In cochannel cells/ sectors (for K = 1 all cells), however, different HSNs are used and the interference diversity effect as described above is fully utilized.

The coding scheme for regular voice transmission in GSM is outlined in Figure 8.4. The information in one digitally encoded speech block, arriving from the speech encoder every 20 ms, is encoded and spread over 8 bursts in 8 subsequent TDMA frames. To achieve a higher degree of diversity, two speech coder blocks are transmitted in parallel in a staggered fashion (one new block transmission starts every 10 ms but takes 20 ms to complete). This means that one code block of 456 bits is spread over 8 half bursts (57 bit).

As one can see the code rate of the code is about 0.5, that is, almost half of the transmitted bits are redundant (check) bits. The coding scheme described in the GSM specifications is theoretically capable of correcting most data frames where 2 bursts (out of 8) are lost due to bad conditions. Most practical implementations do not use straightforward erasure decoding but use so-called "soft" decoding techniques. In these schemes information about the received signal quality gathered either inspecting the analog signal

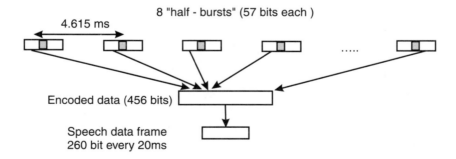

Figure 8.4 Outline of GSM coding and interleaving structure. Each speech frame is encoded into 456 bits that physically transmitted in chunks of 57 bits that occupy one half of 8 bursts which are each transmitted at a different frequency. Two speech frames are transmitted in parallel (in a staggered fashion).

level or by other means (so-called "side-information") is used to locate erroneous bursts to improve error control capabilities [9].

The performance improvement for a GSM system using frequency hopping, compared to a nonhopping system was first demonstrated in [10]. A detailed study of the interference diversity effect can be found in [11]. The graphs in Figure 8.5 illustrates the performance gain as a function of the number of frequencies n. This is an important issue to many GSM operators who have already installed GSM systems without planning for SFH. Using a 3/9 or 4/12 frequency pattern, introducing FH will improve the outage performance of the system, but will hardly improve the capacity since such a system is mainly blocking limited (e.g., suffers from a shortage of channels). In addition, in such a frequency plan the number of channels n is quite low, which further takes some of the edge of the diversity gain

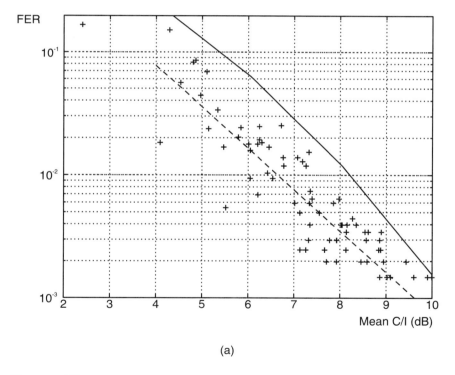

(a)

Figure 8.5 FER as function of average C/I for various system loads and number of frequencies Q [11]: a) 25% system load, 3 frequencies; b) 75% system load, 3 frequencies; c) 25% system load, 12 frequencies; and d) 75% system load, 12 frequencies.

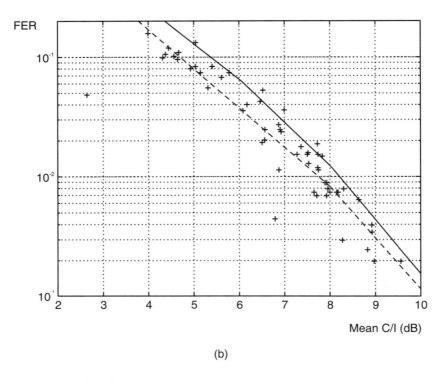

(b)

Figure 8.5 (continued).

(see Figure 8.4). Instead, to get substantial capacity improvements, one has to replan the system and decrease the cluster size. Shown are 1/3 design to give an almost two-fold capacity increase compared to a static assignment scheme.

Problems

*= RUNE simulation solution required

8.1 In a multiuser system, frequency hopping is used as multiple access scheme. In the system, M terminals hop over $Q = 100$ frequencies. In each hop, each terminal (pseudo-) randomly selects two frequencies, denoted f_1 and f_0. If the terminal intends to transmit a 1, frequency f_1 is used. If a 0 is to be transmitted, frequency f_0 will be selected by the terminal. All M users are assumed to be chip synchronized and select frequencies independently of each other. The channel is of OR-type

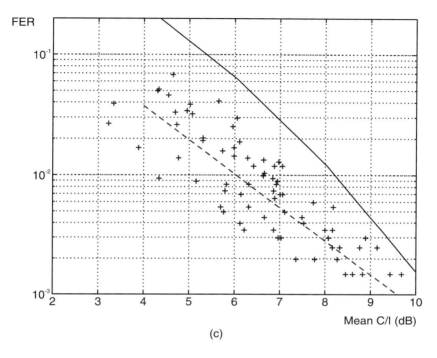

Figure 8.5 (continued).

(i.e. the receiver is only able to detect if there is energy present on the frequency or not). If there is energy present on both f_1 and f_0, the receiver makes a random decision, guessing for a 0 or a 1 with probability 1/2. If there is energy present only on one of the frequencies, the receiver makes the correct decision. The transmitted bits are encoded with a (15,7,5) code. The receiver is equipped with a hard decision decoder, correcting errors up to half the minimum distance. Determine the maximum number of users M that can be admitted in the systems if the bit error probability is not to exceed 5% after decoding!

8.2 A slow frequency hopping system is subject to strong narrowband interference. Assume ideal receivers and that the interfering signal is affecting only one of the N frequencies of the system. For a system with $R = 3/4$ MDS coding with ideal erasure decoding, determine
   a) the raw burst error probability before the decoding,
   b) the symbol error probability after decoding if a random hopping scheme is used,

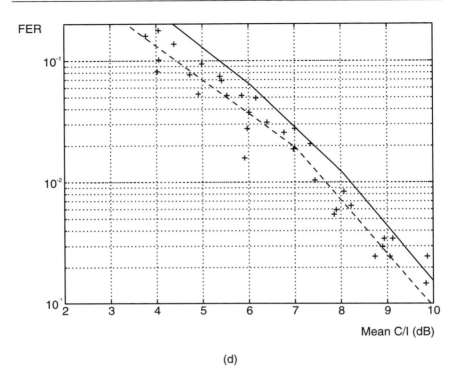

(d)

Figure 8.5 (continued).

c) the symbol error probability after decoding if a cyclic hopping scheme is used.

Neglect other interference sources!

8.3 In a mobile data communication system at 1900 MHz, DPSK modulation is used. The required information rate is 16 kbit/s. A (31,21,5) error correcting block-code is used. The channel can be characterized as a flat Rayleigh fading channel (large number of multipath components) with an average SNR of 20 dB (per info bit). To simplify the analysis we assume that the receiver makes an erroneous bit decision (before decoding) with probability p where

$$p = \{1/2; \text{ if SNR} < 5 \text{ dB}; 0; \text{ if SNR} \geq 5 \text{ dB}$$

a) What is the code word error probability if the mobile receiver is moving with a speed of 20 km/h?

b) Estimate the code word error probability if frequency hopping is used with one hop per transmitted bit! Assume that many frequencies are used and the fading is independent on all frequencies.

8.4 Consider an orthogonal frequency hopping system using a large number of frequencies and omnidirectional antennas and full reuse ($K = 1$). An ideal erasure decoding scheme is used with a long code with rate $R = 3/4$. The hopping sequence for mobiles in different cells can be assumed to be random and uncorrelated. A chip (burst) received with a $C/I > 6$ dB can be assumed error free, otherwise it is declared unreliable and erased. Study the uplink and assume that a constant power control scheme is used, but that the dynamic range (the ratio between the maximum and the minimum transmitter power) of the amplifiers is 20 dB. Estimate the code word error probability as a function of the access port for a mobile in the system!

8.5 Use the assumptions in 8.4, but study the downlink of the system. Compare the results for CRP as in 8.2 and constant transmitter power.

8.6 Consider a frequency hopping system with a frequency plan with 24 carriers arranged in a 1/3 frequency reuse (8 carriers in each frequency group). The frequency hopping sequence for a certain mobile station in a cell is cyclic or random over all frequencies in the frequency group assigned to the cell. In the cyclic hoppping case, all cells use the same hopping sequence. The system is used for digitized voice communication. Each speech frame is interleaved over 8 bursts, and a speech frame is correctly received if at least 6 out of 8 bursts are received with a $C/I > 6$ dB. A user can maximally tolerate a Frame Error Rate (FER) of 5%.

Analyze three frequency hopping cases:

I. Cyclic hopping on frequencies with uncorrelated Rayleigh fading (only frequency diversity).
II. Random hopping on frequencies with uncorrelated Rayleigh fading (both frequency and interferer diversity).
a) Consider the downlink, with stationary users giving 50% channel utilization. Calculate FER for each user and plot the distribution of FER among users in the cases above. Determine which case gives the best speech quality. Try to explain the result and its relation to the receiver model.

b) Plot the FER as function of the system load (channel utilization).
c) Repeat the investigation for the uplink. What are the differences?

Hint:
To simulate on burst level in RUNE, an inner loop can be added. Example for case 2 above:

```
for burst=1:400
  k = rca(b, cpbk);                  % Make a random channel assignment
  Gfading=abs(irandn(size(G)));      % New Rayleigh fading gain matrix
  [..., sirdl_burst, ...]=transmit(...., Gfading.*G, ...);
  sirdl(:,burst)=sirdl_burst; % Collect SIR for all bursts
end
```

References

[1]   Lin, S., and D. J. Costello, *Error Control Coding: Fundamentals and Applications*, Upper Saddle River, NJ: Prentice-Hall, 1983.

[2]   Blahut, R. E., *Theory and Practice of Error Control Codes*, Reading, MA: Addison-Wesley, 1983.

[3]   Andersin, M., and Z. Rosberg, "Time Variant Power Control in Cellular Networks," *Proceedings of PIMRC '96*, Taipei, 1996, pp. 193–197.

[4]   Rosberg, Z., "Fast Power Control in Cellular Networks Based on Short-Term Correlation of Rayleigh Fading," *Proc. Winlab Workshop*, 1997.

[5]   Sunell, K.-E., "Impact of Power Control on Interference Diversity in FH-CDMA Systems," *Proc. IEEE Veh. Tech. Conf. '99*, Amsterdam, Sept. 1999.

[6]   Ahlin, L., and J. Zander, "Principles of Wireless Communication," *Studentlitteratur*, 1997.

[7]   Mouly, M., and M. B. Pautet, *The GSM System for Mobile Communications*, published by the authors, 1992.

[8]   ETSI GSM Technical Specification 05.02, "Multiplexing and Multiple Access on the Radio Path."

[9]   Proakis, J. G., *Digital Communications*, New York: McGraw-Hill, 1995.

[10] Carneheim, C., et al., "Frequency Hopping GSM," *Proc. 44th IEEE Veh. Tech. Conf., VTC '94*, Stockholm, June 1994.

[11] Olofsson, H., J. Näslund, and J. Sköld, "Interference Diversity Gain in Frequency Hopping," *Proc. 45th IEEE Veh. Tech. Conf., VTC '95*, Chicago, July 1995.

[12] Einarsson, G., "Address Schemes for Frequency Hopping Wireless Systems," *Bell Systems Tech J.*, 1981.

[13] Sköld, J., B. Gudmundson, and J. Färjh, "Performance and Characteristics of GSM Based PCS," *Proc. 45th IEEE Veh. Tech. Conf., VTC '95*, Chicago, July 1995.

9

DS-CDMA in Wireless Networks

9.1 DS-CDMA Random Resource Allocation

DS-CDMA has been used in many military applications for secure communications. However, development of the DS-CDMA cellular system was mainly for capacity reasons, in other words, for high spectral efficiency. The purpose of this section is to provide fundamental insights on how the DS-CDMA can increase the radio network capacity.

As was discussed in Chapter 8, the random channel allocation enables each terminal to take its share of the interference (outage problem) of the whole system. Thus the interference for each channel is averaged over all terminals. The random channel allocation potentially improves the system capacity, particularly when it is combined with powerful error correcting codes. The random channel allocation can be achieved at a lower level by a system called Direct Sequence (DS)-CDMA system. Introduced in Chapter 2, a DS-CDMA system uses multiple *access code* in a random fashion rather than changing frequency channels as in the frequency hopping.

This idea is intuitively appealing, but the implementation is not that trivial. An ingenuous way of solving this problem can be found in the North American IS-95 cellular CDMA standard [1]. In this scheme, all terminals use a single and very long m-sequence. In Figure 9.1, the number N denotes the total sequence length, where the sequence is cyclically repeated. Each terminal is identified by a time-offset T_1, T_2 ... in the figure. Terminal 1, for instance will start transmitting its current data symbol using offset T_1. The terminal now picks the $N_{\text{effective}}$ bits following the bit T_1 out of the m-sequence to encode the current information bit. The following information

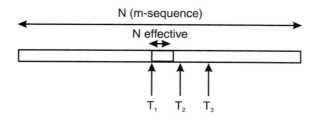

Figure 9.1 Random generation of spreading codes.

bit is then encoded using the sequence bits $T_1 + N_{effective} + 1 \ldots T_1 + 2N_{effective} + 1$ and so on. The bits extracted from the m-sequence basically behave as bits generated by a sequence of fair coin flips. Each terminal will thus in effect use a new random binary sequence of length $N_{effective}$ for every transmitted information bit.

In the IS-95 standard, N is equal to 2^{42-1} or 2^{15-1} and $N_{effective}$ is set to $2^7 = 128$. The number $N = 2^{42-1}$ is used for the uplink, whereas each base station is differentiated by using the m-sequence of a length $N = 2^{15-1}$ in the downlink. For this reason, the uplink m-sequence is often called the *long* sequence, but the downlink sequence is called the *short* sequence. The time-offset is the same as $N_{effective}$, i.e., $T_2 = T_1 + N_{effective} + 1$ and $T_3 = T_2 + N_{effective} + 1$ so on. Therefore, in principle, there are $(2^{42-1})/2^7$ terminals supported by m-sequence. As mentioned in Chapter 2, this way of differentiating users makes the synchronization at the receiver much easier (sliding correlation). Even if this self-orthogonality property is useful for synchronization, the cross correlation property should be sacrificed. Thus, some amount of other user interference at the receiver is unavoidable (see Section 2.1.4).

However, the randomization of the interference is more effective in the DS-CDMA system than in the frequency hopping system. This is due to the fact that channels are swapped for each transmitted symbol, and not for each burst. Therefore with DS-CDMA, the universal frequency reuse ($K = 1$) is applicable to cellular systems. The universal frequency reuse removes the complexity of frequency channel allocation discussed in Chapters 4 and 7. The penalty is that there is no way to keep signals within one cell perfectly orthogonal, that is, free from interference. All other terminals, the terminals both in the cell and in other cells will contribute to the interference experienced by the access port. Since the terminals in the cell are comparatively close, in fact most of the interference power received by the access port will come from terminals within the cell.

DS-CDMA uses more bandwidth (spreading) than what is required to send the information bits. For instance in the IS-95 system, one information bit is encoded by 128 coded bits (chips). Therefore, the natural question is "Is DS-CDMA spectrally efficient?" Obviously, from a single user (signal level) communication point of view, DS-CDMA has no merit in terms of spectrum utilization. However, if it is used in a system that supports multiple users simultaneously, the situation would be different. This question is focused on in the following subsections.

9.2 Simple Capacity Estimation of DS-CDMA Systems

Initially, a simple capacity estimation of a single-cell DS-CDMA system is considered here. The focus is on the uplink capacity, assuming a perfect power control (constantly received power) is achieved at the base station. Assume that there are N active terminals in the cell. The bit energy per interference spectral density ratio, E_b/I_0 of a terminal can be described by:[1]

$$E_b/I_0 = \frac{S/R}{I/W} = \frac{P}{(N-1)P} \cdot \frac{W}{R} = \frac{W}{(N-1)R}$$

where S = received signal power; I = interference power; R = data rate; W = spreading bandwidth (or chip rate); P = constant received power from a terminal.

The above equation can be rewritten in the following form:

$$N = 1 + \frac{W/R}{E_b/I_0}$$

Therefore the number of supportable terminals in the cell is proportional to the processing gain W/R, and is inversely proportional to the target E_b/I_0 at the base station receiver. In the IS-95 system, the data rate is 9.6 kbps and the processing gain is 128 (about 21 dB, $W = 1.25$ MHz), and the target E_b/I_0 for achieving a BER 10^{-3} (standard for voice communication systems) corresponds to 7 dB [2]. With this setting, 26 users may be supported by the cell when the constant received power control is applied:

$$N = 1 + \frac{W/R}{E_b/I_0} = 1 + \frac{128}{5.01} = 26$$

---

1. In conventional systems, E_b/N_0 is used, where N_0 is the noise power density. However for the DS-CDMA system, since the dominating factor is interference from the other cochannel users, the notation I_0 is employed.

In the same manner, it is clear that N ranges from 17 to 65 users when the target E_b/I_0 varies from 9 down to 3 dB. If the same bandwidth 1.25 MHz is used for the FDMA based AMPS [3], then there will be 42 users since AMPS uses a 30 KHz channel for each user (1.25MHz/30KHz).

In fact, there is no significant gain in terms of capacity increase from DS-CDMA. Now, consider the multiple-cell system. To begin with, define the following frequency reuse factor

$$F = \frac{\text{Total interference power}}{\text{Own cell interference power}}$$

Using this definition, the received E_b/I_0 in the multiple-cell system can be derived

$$E_b/I_0 = \frac{S/R}{I/W} = \frac{P}{F(N-1)P} \cdot \frac{W}{R}$$

This gives the capacity form in the multiple-cell system as

$$N = 1 + \frac{W/R}{(E_b/I_0)F} \tag{9.1}$$

The analytic study shows that the fraction of interference that comes from the neighboring cells is about 50–60% of its own cell interference, when the users are uniformly distributed over the system [2], [4]. For example, in the IS-95 system, assuming $F = 1.6$ and $E_b/I_0 = 7$ dB, there are 16 users per cell, whereas AMPS can support 6 = (1.25 Mhz/30 KHz)/7 users per cell when the cluster size (K) is 7. When E_b/I_0 is varying from 3dB to 9dB, the number of users per cell will also change, ranging from 41 to 11. Therefore, as moving to the multiple-cell system, the capacity gain from DS-CDMA system becomes considerable. The randomization of the interference and interference rejection in DS-CDMA enables universal frequency reuse $(K = 1)$. This in turn results in capacity difference between random (DS-CDMA) and static (FDMA) channel systems.

Now revisit the capacity estimation (9.1). The advantage in terms of capacity increase from using the wider bandwidth W is twofold. As the bandwidth becomes wide, the processing gain is improved, potentially rejecting a large amount of interference. At the same time, there is much room for introducing powerful error correcting codes with high redundancy. Thus, it is possible to have a much lower E_b/I_0 target, while keeping the bit error rate constant. However, since expanding spreading bandwidth is an expensive procedure in many practical situations, the capacity improvement should be carefully estimated.

It is clear from the factor F in (9.1) that the DS-CDMA system's capacity is rather flexible depending on the traffic distribution. For instance, if the outer cell's interference becomes smaller (small F), then the capacity of a given cell will be increasing. So, in a sense, the capacity is not fixed, and is rather said to be *soft*.[2]

Example 9.1 Capacity Estimation of a DS-CDMA System

A DS-CDMA system for mobile telephony operates with an effective sequence length of 100 chips. The propagation conditions are characterized by a distance dependent power-law path loss ($\alpha = 4$). The system uses convolutional coding such that the received E_b/I_0 should not be less than 7 dB in a link in order to provide a sufficient bit error performance. The mobiles are assumed to be rather many and uniformly distributed over the cell. The mobiles use perfect power control, that is, identical power level is received from each mobile at the access port. Assume that 50% of the total interference power emanates from the own cell. Estimate the maximum number of mobiles that may be active at any time in the cell in order to satisfy the SIR requirement for all mobiles. What happens if this number is exceeded?

Solution:

For using the capacity (9.1), first specify the numbers as below:

$$W/R = 100$$
$$E_b/I_0 = 7 \text{ (dB)}$$
$$F = 2$$

---

2. Regarding this aspect, it will be worthwhile to quote the following from [4]: "In CDMA, frequency reuse efficiency is determined by the SIR resulted form all the users within the range, instead of users in any given cell. Since the total capacity becomes quite large, statistics of all the users are more important than those of a single user. The low of large numbers can be said to apply. This means that the net interference to any given signal is the average of all users' received power times the number of users. As long as the ratio of received signal power to average interference power density is greater than a threshold, the channel will provide an acceptable signal quality. With an FDMA or TDMA system, interference is governed by a law of small numbers in which the worst-case situations determine the percentage of time in which the desired signal quality will not be achieved."

Then,

$$N \approx 1 + \frac{100}{5 \cdot 2} = 11$$

Note that exceeding this limit will drop the SIR below the required level for all mobiles, supporting no mobile.

9.3 Refined Capacity Estimation of DS-CDMA Systems

Due to the universal frequency reuse ($K = 1$), sectorization (e.g., typical 120 degree sector antennas) in a DS-CDMA system has the same effect as cell splitting (microcellular system). That is, the CDMA system's capacity is linearly proportional to the number of sectors per cell, even if sectorization will cause more frequent intracell handoffs (softer handoff). On the other hand, sectorization gain in FDMA or TDMA systems requires careful frequency planning. Otherwise, the capacity improvement may not be large enough compared to that of DS-CDMA systems.

Another factor defining DS-CDMA system's capacity is the use of instantaneous rate selection. One way of increasing the data rate in a DS-CDMA system, while using the same bandwidth is to use a high transmission power to cope with the decreased processing gain. In other words, terminals can decrease transmission power when their rates are reduced. Decreasing transmission power will in turn decrease the total interference amount in the system, and other users can gain advantage from the lowered interference situation. This is a smooth way of shifting one user's unused capacity to the other users in need. A similar method in the F/TDMA based system is the Dynamic Channel Allocation (DCA) discussed in Chapter 7. But, DCA requires a high signaling burden and increased hardware complexity in many cases.

A smart way of utilizing the instantaneous rate selection can be found in IS-95. Experimental research shows that only 35–40 percentage of time is active (nonzero) in human speech conversation. In IS-95 system, human speech is encoded into four data rates, 9.6, 4.8, 2.4, and 1.2 Kbps. When speech users are in the silent mode, the date rate will be encoded into 1.2 Kbps. Therefore, one user's unused capacity (silence) can be instantaneously utilized by other users in need. It has been reported that almost two times more capacity could be achieved by utilizing voice activity defection when the human voice activity factor is 40% [4]. As a result, the capacity equation is refined to

$$N = 1 + \frac{W/R}{(E_b/I_0)F} \cdot S \cdot D$$

where S and D denote the sectorization gain and data rate selection gain, respectively.

Assuming $E_b/I_0 = 7$ dB, $S = 3$ (3 sectored per cell) and $D = 2$, there are 91 users per cell in the IS-95 system. Thus the number of users per cell is more than 15 times larger than that of AMPS (6 = (1.25 Mhz/ 30 KHz)/7).

9.4 Erlang Capacity of DS-CDMA Systems

When cellular CDMA was introduced in the early 1990s, the main argument was that the capacity of a DS-CDMA system could be more than ten times larger than that of AMPS. So far, this proposition has been examined through a simple capacity calculation. Due to [5], however, the capacity derivation can be refined into a more advanced form.

The refinement will start with the uplink in a single cell. Assume that users will arrive at the cell according to the Poisson process with rate λ and each user's service time will follow the exponential distribution with mean $1/\mu$. Under this setting, it can easily be seen that the number of users per cell, say k_u, will follow the Poisson distribution with mean λ/μ in the steady state.

At a given instance in time, each user may be active or not, which is assumed to have the following distribution.

$$\rho = \Pr(\nu_i = 1) = 1 - \Pr(\nu_i = 0)$$

where ν_i denotes the user activity status with one being active; zero being nonactive. Furthermore, the constant received power is assumed. Then the total received power at the base station is described by

$$\sum_{i=1}^{k_u} \nu_i E_b R + N_0 W$$

where N_0 is the density of noise power at the receiver and the other symbols are the same as defined earlier in this chapter.

Now focus on the user with index 1. Interference plus noise power for this user can be described by

$$I_0 W = \sum_{i=2}^{k_u} \nu_i E_b R + N_0 W$$

In the power controlled DS-CDMA system, as the number of users is increasing, the interference will also increase. However, if the interference amount is above a certain threshold, then power value for each user will increase very rapidly to cope with the large interference, leading to the so-called cocktail effect. Therefore, the system becomes extremely unstable. In principle, the DS-CDMA system has no limitation in accepting new users (no hard blocking). However, if the system becomes congested, none of the users can communicate properly due to increased interference with fast power control. Therefore, a system stability threshold η can be defined in a way that the relative value of the total interference power compared to receiver noise should satisfy

$$\frac{I_0 W}{N_0 W} \leq \frac{1}{\eta}$$

For instance, when η is 0.1, the system is said to be congested if the total interference power experienced by a single user is ten times larger than the receiver noise.

If the interference situation does not satisfy the above condition, then the system becomes unstable. Rewriting this condition will give

$$\sum_{i=2}^{k_u} \nu_i \leq \frac{(W/R)(1-\eta)}{E_b/I_0}$$

Therefore the important question at hand is the probability given by

$$
\begin{aligned}
P_{\text{unstable}} &= \Pr\left(\sum_{i=2}^{k_u} \nu_i > \frac{(W/R)(1-\eta)}{E_b/I_0} \right) \\
&\cong \Pr\left(\sum_{i=1}^{k_u} \nu_i > \frac{(W/R)(1-\eta)}{E_b/I_0} \right) \qquad (9.2) \\
&= \Pr\left(Z > \frac{(W/R)(1-\eta)}{E_b/I_0} \right)
\end{aligned}
$$

It can be proved that the random variable Z is Poisson distributed with mean $\rho(\lambda/\mu)$ when k_u follows the Poisson distribution with mean

λ/μ. Thus, the probability (of system instability) can be calculated for a given ρ and λ/μ. In other words, the maximum load λ/μ can be derived, for a given system instability probability, when the user activity factor ρ is fixed. This is very similar to the conventional Erlang capacity calculation where the maximum load λ/μ that will give the blocking probability less than a certain value (e.g., 2%) is derived, when the number of channels per cell is given by K_0.

$$P_B = P_{K_0} = \frac{(\lambda/\mu)^{K_0}/K_0!}{\sum_{j=0}^{K_0} (\lambda/\mu)^j/j!} \tag{9.3}$$

Capacity derivation in the multiple-cell DS-CDMA system can be easily extended from the single-cell case by considering the multiple cells as an *aggregated* single cell. The only change is to replace the mean of Z in (9.2) by $\rho(1 + f)(\lambda/\mu)$, where the number f is the fraction of interference coming from the outer cells compared to the own cell. Therefore, using (9.2), the system instability probability in the multiple-cell case holds as follows

$$P_{\text{unstable}} \approx e^{-\rho(1+f)(\lambda/\mu)} \sum_{k=K_0}^{\infty} (\rho(1+f)(\lambda/\mu))^k/k! \tag{9.4}$$

where $K_0 = \left\lfloor \dfrac{(W/R)(1-\eta)}{E_b/I_0} \right\rfloor$ is the largest integer that does not exceed $\dfrac{(W/R)(1-\eta)}{E_b/I_0}$

Example 9.2 System Unstable Probability Calculation

Consider the uplink of a DS-CDMA system, where the processing gain is 20 dB. The number of users in the cell is Poisson distributed with the mean of λ/μ. For a given instant, the probability that a randomly chosen user is active is 0.4. Target E_b/I_0 is 7 dB. The system conducts the constant received power control. The other cell interference is 60% of own cell interference. Define that the system becomes unstable if the interference is 10 times greater than the background noise amount. Determine the maximum load (capacity) per cell that gives the system unstable probability less than 2%.

Solution:

Using (9.4) with $K_0 = \left\lfloor \dfrac{10^2 (1 - 0.1)}{7 \text{ dB}} \right\rfloor$ and $f = 0.6$ the probability can be calculated by

$$P_{\text{unstable}} \approx e^{-0.4 \cdot 1.6 \cdot (\lambda/\mu)} \sum_{k=17}^{\infty} (0.4 \cdot 1.6 \cdot (\lambda/\mu))^k / k!$$

for a given λ/μ. Figure 9.2 shows probabilities for different λ/μ settings. From the figure, it can be seen that $\lambda/\mu = 15$ is the maximum load per cell.

9.5 Capacity of Multi-Service DS-CDMA Systems

So far, the discussion has been focused on the single-rate, voice communication system. The next-generation DS-CDMA systems will be able to provide multimedia services. Such services include package-switched connections transporting emails, files, WWW pages, video, and so forth. Those services are characterized by different QoS requirements such as transmission and error rates. Therefore, the capacity of a system can be defined in terms of number of users whose QoS requirements have been satisfied.

To investigate this problem further, consider the uplink a single cell, where N terminals are using the same frequency channel. Assume a short time interval such that the link gain between each terminal i and the base station is stationary and given by g_i. Given a power vector $P = (p_1, p_2, \ldots p_N)$, the received SIR of terminal i, is defined by

$$\Gamma_i = \frac{g_i p_i}{\displaystyle\sum_{j \neq i} g_j p_j + n + I_{\text{inter}}}$$

where n is the noise power at the base station and I_{inter} is the intercell interference power from other cells.

Suppose that terminal i can utilize a number of discrete transmission rates $r_i = \{r^1, r^2, \ldots, r^K\}$ by varying its processing gain. Then, the E_b/I_0 of the terminal can be described by

$$(E_b/I_0)_i = \frac{W}{r_i} \cdot \Gamma_i$$

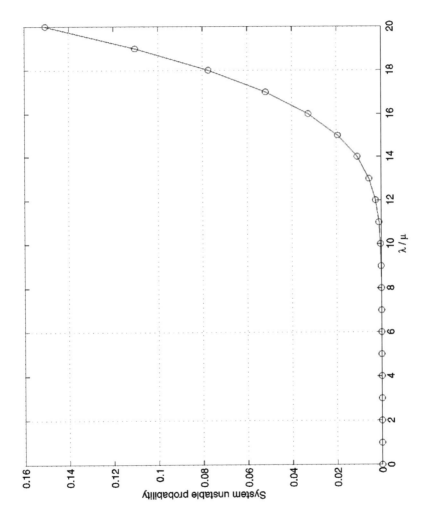

Figure 9.2 System unstable probability under different cell loads.

The realized data rate of a termnal depends on many factors such as receiver structure, user mobility, etc. Nevertheless, SIR is a reasonable measure to match effective data rates. To receive information properly at a transmission rate r^k with a certain error rate, terminal i is expected to attain an SIR, not less than γ^k. Let γ_i be the target SIR of terminal i such that $\gamma_i \in \{\gamma^1, \gamma^2, \ldots, \gamma^K\}$ where each target element corresponds to each rate, respectively.

With a certain assignment of SIR targets to terminals, the power value that supports every terminal with the minimum power can be found by solving the linear equation system:

$$\Gamma_i = \frac{g_i p_i}{\sum_{j \neq i} g_j p_j + n + I_{\text{inter}}} = \gamma_i, \ \forall \ i$$

If the above linear equation system has a solution, the power value of terminal i can be proved to be:

$$p_i = \frac{n + I_{\text{inter}}}{1 - \sum_{j=1}^{N} \dfrac{\gamma_j}{1 + \gamma_j}} \cdot \frac{\gamma_i}{g_i(1 + \gamma_i)}$$

Since the power value should be positive (assuming positive receiver noise and the other cell interference), the following *necessary* condition for supporting every terminal with its target rate (SIR) can be derived.

Proposition 9.1

If the target rate (SIR) of each termnal is achieved, then

$$\sum_{j=1}^{N} \frac{\gamma_j}{1 + \gamma_j} < 1$$

Example 9.3 Uplink Capacity in a Multi-Service DS-CDMA System

Consider the uplink of a single cell in a DS-CDMA system. The system uses a 5 MHz bandwidth, where terminals are transmitting with a data rate $R = 10$ Kbps or 20 Kbps. All terminals need a minimum $E_b/I_0 = 7$ dB (≈ 5), regardless of data rates. There is no limitation on the terminal's transmission power. Assume that half of the active users require 10 Kbps and the outer-cell interference is fixed. Determine the maximum supportable number of active terminals in the cell.

Solution:

First assume that the number of active users is $N = 2M$. Then, derive the target SIRs for 10 Kbps and 20 Kbps users by

$$\gamma_1 = (E_b/I_0) \cdot (r_1/W) \approx 10^{-2}$$

$$\gamma_2 = (E_b/I_0) \cdot (r_1/W) \approx 2 \cdot 10^{-2}$$

Using Proposition 9.1:

$$\sum_{i=1}^{M} \frac{\gamma_1}{1+\gamma_1} + \sum_{i=1+1}^{2M} \frac{\gamma_2}{1+\gamma_2} = M\left(\frac{10^{-2}}{1+10^{-2}} + \frac{2 \cdot 10^{-2}}{1+2 \cdot 10^{-2}}\right) < 1$$

$$M < \left(\frac{10^{-2}}{1+10^{-2}} + \frac{2 \cdot 10^{-2}}{1+2 \cdot 10^{-2}}\right)^{-1} = 33.89$$

$$N = 2M < 67.78$$

Therefore, the maximum number of supportable users is 67.

Now consider that each terminal i is power-limited by $\overline{p_i}$, then

$$0 < p_i = \frac{n + I_{\text{inter}}}{1 - \sum_{j=1}^{N} \frac{\gamma_j}{1+\gamma_j}} \cdot \frac{\gamma_i}{g_i(1+\gamma_i)} \le \overline{p}_i, \ \forall \ i$$

It is easy to show that this is equivalent to

$$\sum_{j=1}^{N} \frac{\gamma_j}{1+\gamma_j} \le 1 - \frac{n + I_{\text{inter}}}{\min_{1 \le i \le N} \left[\overline{p}_i g_i \frac{1+\gamma_i}{\gamma_i}\right]}$$

Proposition 9.2 (Power-constrained case)

If the target rate (SIR) of each terminal is achieved within each terminal's power limit, $\overline{p}_i$ then

$$\sum_{j=1}^{N} \frac{\gamma_j}{1 + \gamma_j} \leq 1 - \frac{n + I_{\text{inter}}}{\min\limits_{1 \leq i \leq N}\left[\overline{p}_i g_i \dfrac{1 + \gamma_i}{\gamma_i}\right]}$$

To maximize the number of supported users per cell, QoS requirement of each service should be exploited. For instance, the radio network will try to support voice/visual communication in a real time manner. However, those services such as file transfer can endure a certain amount of delay, and the radio network can fully utilize their latency tolerance to improve the spectrum efficiently. In other words, delay-insensitive users can be supported in a best effort fashion in the sense that they may use any excess capacity that the radio network can provide.

If a user would like to send data with a high transmission rate, then a high transmission power might be needed to cope with the decreased processing gain, interfering with the other users. In particular, if the user has an unfavorable propagation condition, he or she cannot increase his or her rate significantly without damaging the other users severely. Thus, if that user is not delay-sensitive, it is desirable to reduce his or her data rate, even to zero. Setting transmitter powers to zero, can be denoted as a hybrid of TDMA and CDMA. It was shown in [9] that under certain conditions inducing TDMA on a CDMA system, the total energy consumption could be decreased while offering the same throughput, compared to pure CDMA. Further, we can prove the following property [6].

Proposition 9.3[3]

Consider the uplink of a single cell in a DS-CDMA system. Denote each terminal in the system by 1, 2, ... N. Further, assume that the link gain g_i of each mobile i, satisfies $g_1 > g >, \ldots > g_N$ and that there exists a feasible target assignment that can support every terminal with the targets. Among the feasible reassignments of those targets, the total power sum of the terminal is minimized by reassigning the target SIRs such that $\gamma_i > \gamma_2 >, \ldots > \gamma_N$.

The theorem says that the total power of the mobiles can be reduced while maintaining the same throughput of the bare station, by reassigning target SIRs in the decreasing manner, according to the user link condition. The main idea behind this is that data rates (and thus power values) of users with unfavorable link conditions are decreased even to zero, while the users with high link gains transmit high rate data with the maximum power.

---

3. This is the same as Proposition 1 in [6].

Unless it does not damage user QoS severely, the users can minimize their energy consumption and the system can maximize the throughput by postponing transmission until the link conditions become favorable.

9.6 Power Control in DS-CDMA Systems

As has been noted in previous chapters, SIR exhibits considerable fluctuations caused by propagation condition variations as the mobile travels around the service area. In particular under Rayleigh fading environments, the phenomenon will be much more serious. Furthermore, in the uplink side, there will be a big difference in path gains among the terminals. If all the mobiles were to use the same transmission power, the user's signal power close to the base station will jam the far user's signal. This problem is called the near-far problem.

In the previous sections, a perfect power control (constant received power) was assumed to derive DS-CDMA system's capacity. Therefore the results in those sections represent an upper bound on the DS-CDMA system's capacity. For instance in Example 9.1, it is possible for the system to support only one terminal if every terminal transmits with the same power and the closest terminal masks out the other terminals. Figure 9.3 shows the effect of imperfect power control in terms of outage probability. The power control has log-normally distributed errors with standard deviations of 3 and 5 dB, respectively. The reference perfect power control algorithm was the constant received power control, that is, the standard deviation of errors is 0 dB. At the quality of service (QoS) level of 10% outage probability, the perfect power control can support more than three times the number of users, compared to the imperfect power control with the 5dB error.

Efficient power controls are needed in practical DS-CDMA systems to cope with the near-far problem and fading phenomenon. This is one of the critical issues in using DS-CDMA technique to support simultaneous users, not a single signal level communication. Therefore, a question at hand is if it is possible to implement any perfect power control in a real DS-CDMA system.

Various aspects on the power control have been discussed in Chapter 6. Implementing the power control algorithms described in that chapter to a real system, however, requires some modification. The reasons can be explained by using the DPC algorithm described in Section 6.3.1.

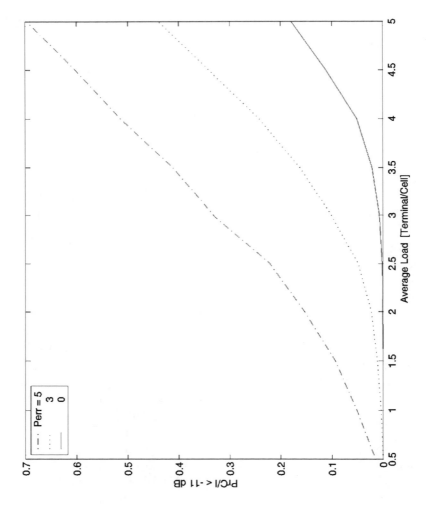

Figure 9.3 Imperfect power control and DS-CDMA system capacity.

$$P_i^{(n+1)} = \frac{\gamma_0}{\gamma_i^{(n)}} P_i^{(n)}$$

In order to implement DPC, in principle, transmission powers should be increased or decreased quickly to follow the track of DPC. However, in real implementation, it is rather difficult to increase (or decrease) transmission power quickly and unlimitedly. Secondly, besides the efforts on measuring $\gamma_i^{(n)}$, there will be some measurement errors. Even if the measurement is relatively reliable, delivered information to the transmitter will be a delayed version of the original one due to the distance between the transmitter and the receiver. Thirdly, since the fluctuation in $\gamma_i^{(n)}$ is large, it requires a large signaling bandwidth to deliver the exact information about $\gamma_i^{(n)}$ to the transmitter.

In order to cope with practical limitations, a slightly different approach is used in the practical DS-CDMA systems. There are basically two types of power control algorithms in real systems, called *open loop* and *closed loop* power controls. Open loop power control does not require any feedback measurement on the transmitter signal quality at the receiver. This kind of power control is needed for the users whose data size is very short and thus it does not have enough time to measure the signal quality at the receiver. In this kind of category, it refers to wireless packet access with a small packet size, where instantaneous initial power calculation is needed. For the purpose, the transmitter will usually measure the signal strength of the other direction's radio link. For example, the terminal will measure the signal strength from the access port side. In the IS-95 system, this is accomplished by utilizing the automatic gain control (AGC) circuit of the receiver. The AGC circuits operate on the receiver's IF frequency amplifier so that the input to receiver's analog-digital (A/D) converters is held constant. The AGC control is used to control the gain of the transmitter IF amplifiers exactly in step with the receiver's IF gain. Thus, if the mobile terminal moves closer to the base station, increasing the received signal level, the receiver AGC will reduce the receiver IF amplifier gain, and the transmitter IF gain. Due to the lack of reciprocity in radio channels, however, this kind of power control may not be accurate.

Due to the frequency separation, uplink and downlink channels will experience completely different radio propagation conditions. In particular when the (uncorrelated) Rayleigh fading results in sharp decreases in the received signal, the open loop power control can make things worse by increasing/decreasing transmission power rapidly. To take care of this undesirable phenomenon, the closed loop power control is considered. Closed loop power control requires a feedback measurement as in DPC to adjust transmitter powers. However, due to limitations cited above, a simplified version is used in practical systems. That is, the receiver measures the signal quality (SIR) of the transmitter and compares it with a target threshold. If the received quality is less than the target, the receiver will send a binary power-up command to the transmitter. Otherwise, a power-down command will be delivered to the transmitter. If the measurement occurs N times per second, then it needs the power control channel with the speed of N bps. Upon receiving the power control bits, either up or down, the transmitter will increase or decrease its power by the fixed amount, that is, 1 dB up or 1 dB down. In the uplink of the IS-95 system, 800 bps power control speed was suggested. That is, SIR measurement is made every 1.25 ms, giving an 800 bps power control command stream in the downlink. The value represented by the closed loop power control is converted to an analog voltage and then added to the open loop control voltage (the receiver AGC signal), and applied to the transmitter gain control circuits. In the downlink of the IS-95 system, however, a slightly different and slow control mechanism is used for the closed loop power control. Figure 9.4 contains a simulation result of IS-95 uplink power control. It shows the received SIR at the base station receiver in every 1.25 ms. The target E_b/I_0 is set to 3 dB and the figure shows the combined effect of closed and open loop power control. The received E_b/I_0 is centered to the target with some deviations. It has been demonstrated in [8] that inaccuracy of the IS-95 uplink power control is approximately log normally distributed with a standard deviation between 1 and 2 dB.

In the third generation system, such as ETSI, WCDMA [7] will have a similar closed power control as in the uplink of the IS-95 but with a two times higher speed, 1600 times per second. Unlike to the IS-95, this kind of power control is applied to both uplink and downlink. Figure 9.5 shows the physical channel structure of WCDMA. The channel is logically composed of data and control channels, called DPDCH and DPCCH, respectively. In the downlink, each frame is represented by 10-ms data amount, which is further divided into 16 slots. Each time slot contains data and control fields (DPDCH and DPCCH). The power control command (bit) will be inserted in the TPC filed in the DPCCH part of the slot in every 0.625 ms. Note that increasing power control speed would improve the power control efficiency. However, at the same time, it will increase the measurement and signaling burden.

In the closed loop power control, the target SIR values are controlled by another loop, called outer loop power control. The outer loop measures

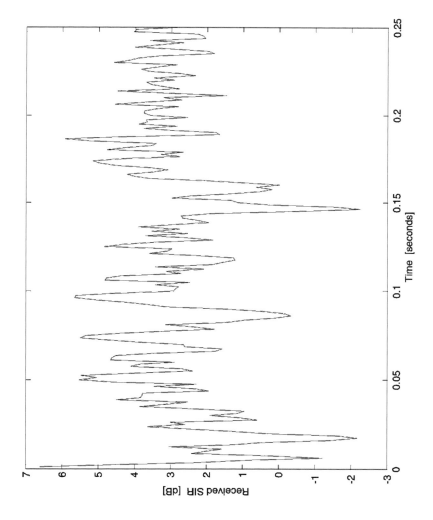

Figure 9.4 Uplink link power control in IS-95 system.

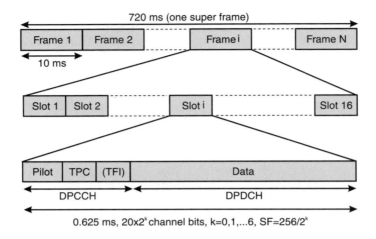

Figure 9.5 WCDMA physical downlink [7].

the link quality, typically frame error rates and adjusts the SIR targets accordingly. Ensuring lowest possible SIR target is used at all times to maximize the network capacity. Outer loop power control is used in both IS-95 and WCDMA systems.

9.7 Soft Handoff in DS-CDMA Systems

Handoff is one of the key aspects that supports user mobility in wireless networks. During handoff, a mobile user will be connected to the new access port as he or she approaches it. Soft handoff means that a mobile is connected to more than one base station, that is, the current and potential new base station for a short time until the terminal is fully supported by the target base station. Depending on the number of base stations involved, 2-way or 3-way soft handoffs will occur.

Contrary to the soft handoff, conventional hard handoff is characterized as *break-before-make*. With hard handoff, there is an instant during which no base station supports the terminal. The main drawback of hard handoff is in the so-called ping-pong phenomenon, causing very frequent switches between base stations. This undesirable situation is problematic especially when the terminal is moving parallel to cell borders. Handoff is one of the procedures that impose a large amount of signaling on the system. Therefore, the ping-pong problem causes an extremely heavy burden to the system. To avoid this situation, some threshold margin (hysteresis) may be used in

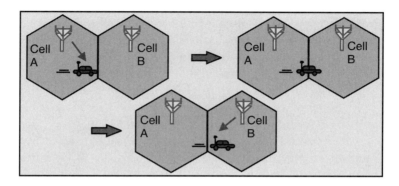

Figure 9.6 Hard handoff (break-before-make).

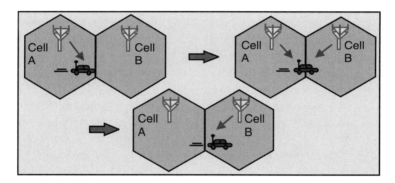

Figure 9.7 Hard handoff (make-before-break).

determining (executing) handoff procedures. That is, the new base station is connected only after the signal strength from the target base station is sufficiently large enough, compared to the current base station's signal strength. However, this will increase handoff failure probability by unnecessarily delaying the handoff procedure. Furthermore in DS-CDMA systems, if the old station leans toward keeping the current connection, then the power control rapidly increases transmission powers for the outgoing (moving-out) terminals, and the system will experience a higher level of interference.

Soft handoff may completely remove the ping-pong phenomenon. Furthermore in DS-CDMA systems, by employing soft handoff, terminals can reduce its transmission power and the total interference of the system can be decreased. When a terminal is under soft handoff, it receives the power control commands from different base stations. It is possible that one base station sends power-down commands, whereas the other base station sends power-up commands. This happens most probably when the terminal is approaching a new base station. In this situation, the terminal will decrease its power since it is possible to communicate with a reduced transmission power. This means that the terminal follows the power control command from the best base station in terms of giving the best signal strength report.

In the IS-95 system, the soft handoff state of a terminal is based on the pilot strength measurement. The set called *active* contains the identifications of the base stations that are currently communicating with the terminal. Therefore if the number of base stations in the active set is more than one, it means the terminal is undergoing soft handoff. As in Figure 9.8, there are basically three thresholds that determine the soft handoff status of a terminal, T_add, T_drop and T_tdrop. If the received E_c/I_0 (chip energy per interference spectral density) of a pilot channel of a base station in the active set is below T_drop and stays there until a timer expires (T_tdrop), then the connection to the base station is removed. If the received E_c/I_0 of a pilot channel currently not belonging to the active set is above T_add, then the pilot channel is moved to the active set, that is, a new connection is established.

In IS-95 soft handoff, if both thresholds T_add and T_drop are increased simultaneously, then the soft handoff region will be decreased, whereas if they are decreased simultaneously, then soft handoff region is increased. Figure 9.9 show a simulation result on IS-95 CDMA soft handoff. The figure contains soft handoff rates, ratios of the number of terminals under soft handoff to the number of active terminals at a given instant. In

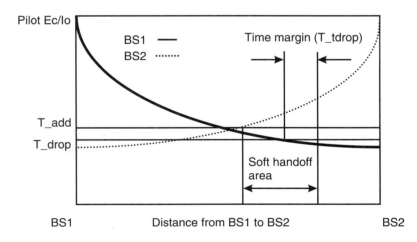

Figure 9.8 Soft handoff in IS-95 system.

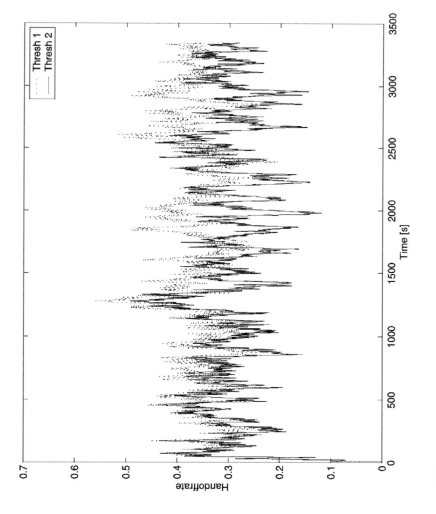

Figure 9.9 Soft handoff thresholds and soft handoff rates in IS-95 system.

this figure, most of the time soft handoff rates are in the range of [0.25, 0.45]. By shifting handoff parameters from T_add = −14 dB and T_drop = −16 dB (Thresh1) to T_add = −12dB and T_drop = −13 dB (Thresh 2), the handoff rate is obviously decreased.

The same phenomenon will happen if the E_c/I_0 from both base stations is either decreased or increased, without changing the thresholds. Thus, if the interference amount in the downlink side is decreased (increased E_c/I_0), that is, light traffic load, then the soft handoff region will be enlarged. Therefore handoff rate is changing according to the traffic density. For example, at midnight, handoff rate (region) is increasing whereas during the busy hours the handoff region shrinks.

To avoid such a phenomenon, a slightly different scheme has been considered in WCDMA. When the incoming base station's beacon channel power is larger than the current base station by T_add then a new base station is connected and soft handoff starts. Similarly, when the current base station's beacon channel power is less than that of the incoming base station by T_drop, the connection to the current base station is removed. It can easily be seen from Figure 9.10 that the soft handoff region is invariable to the traffic load.

9.8 Dynamic Cell Management in DS-CDMA Systems

In a DS-CDMA system, the average SIR in a cell decreases as the number of users in the cell increases. Cells under heavy traffic loads experience high

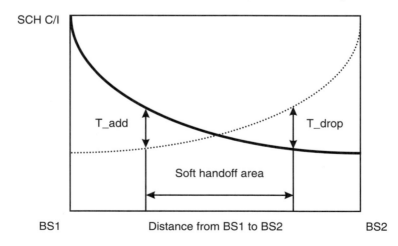

Figure 9.10 Soft handoff in ETSI WCDMA system [7].

interference, where some of the new or ongoing calls would be blocked or dropped as a result of the SIR decrease. On the other hand, neighboring cells under light traffic loads are still able to accommodate more calls. In the DS-CDMA system, the traffic unbalance can be handled autonomously, by the so-called cell breathing. To illustrate the phenomenon, consider Figure 9.11. In this figure, the dotted line denotes the original cell boundary, where it can be seen that there is some level of traffic unbalance between two cells. However, after cell breathing (solid line), the cell with a light load enlarges its size, whereas the cell with a high load shrinks, and thus each cell now has the same amount of traffic load. This is similar to borrowing some channels from the neighboring cells in the DCA schemes. The question is "How can such a phenomenon occur autonomously?"

In the IS-95 based CDMA system, for example, the soft handoff states of a terminal are determined by the pilot strength measurement and the predefined soft handoff thresholds. In Figure 9.8, if the cell load of BS1 is high, then the interference amount in the cell will also increase. As a result, E_c/I_0 from BS1 will be decreasing, which in turn hands over some of the current users to BS2. At the same time, if the traffic amount of BS2 is low, the E_c/I_0 of the pilot from BS2 will be increasing, and there is a possibility of accepting new users from the BS1.

Dynamic cell management (DCM) concept refers to the scheme of activating the cell breathing further, resolving the local traffic unbalance in DS-CDMA systems. With DCM, a heavily loaded cell dynamically shrinks its coverage area to adjust loads it can adequately handle, while each adjacent cell that is less loaded increases its coverage area to pick up the extra traffic.

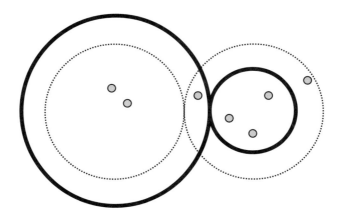

Figure 9.11 Cell breathing in a CDMA system.

There are a few DCM algorithms in literature, which adaptively control the cell coverage by means of dynamically adjusting soft handoff parameters or pilot channel power [10].

Example 9.4 Dynamic Cell Management

In a DS-CDMA system, a traffic controlled handoff scheme is used. The base station varies the power level on their control (beacon) channel and the terminals always connect to the base station that gives received with the highest power. Consider the highway system below with 3 base stations at $x = \{1, 3, 5\}$ km (see Figure 9.12). Assume that the mobiles are Poisson distributed with intensity

$$\omega(x) = \begin{cases} x & 0 \leq x < 2 \\ 3 - x/2 & 2 \leq x \leq 6 \\ 0 & otherwise \end{cases}$$

Assume that the received power is proportional to the 4th power of the distance and that the fading can be neglected. What should the relation between the transmitter powers be in the three base stations such that the cell load will be the same in all base stations?

Solution:

Since the received power is only distance dependent (no fading) there will be distinct handoff region delimited by the points x_1 and x_2 (see Figure 9.13). The expected load is

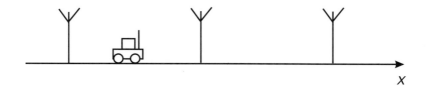

Figure 9.12 Highway cellular system (Example 9.4).

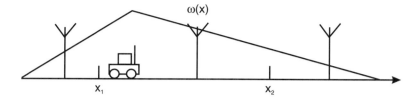

Figure 9.13 Traffic distribution (Example 9.4).

$$\lambda_1 = \int_0^{x_1} \omega(x)dx = \frac{x_1^2}{2} = \lambda_2 = \int_{x_1}^{x_2} \omega(x)dx = 6 - \frac{x_1^2}{2} - \frac{(6-x_2)^2}{4}$$

$$= \lambda_3 = \int_{x_2}^{6} \omega(x)dx = \frac{(6-x_2)^2}{4} \Rightarrow$$

$$x_1 = 2$$
$$x_2 = 6 - 2\sqrt{2} \approx 3.17$$

Now adjusting the powers such that the received power at the handoff points is the same for adjacent base stations yields

$$P_{rx,1} = c\frac{P_1}{(x_1-1)^4} = c\frac{P_2}{(3-x_1)^4}$$

$$P_{rx,2} = c\frac{P_2}{(x_2-3)^4} = c\frac{P_3}{(5-x_2)^4} \Rightarrow$$

$$P_2 = P_1$$

$$P_3 = P_1\left(\frac{2\sqrt{2}-1}{3-2\sqrt{2}}\right)^4 \approx 1.3 \cdot 10^4 P_1 \ (\approx 41 \text{ dB higher than } P_1 \text{ and } P_2)$$

Problems

9.1 In the direct sequence system in Example 9.1, results showed that the number of mobiles per cell could not be larger than 11 in order to make it possible to satisfy the SIR requirements. In a real system, the number of interfering terminals is not fixed, but can be seen as a random number. Let us investigate the case where the number of terminals in a cell has a Poisson distribution. For the sake of simplicity, assume, however, that the interference emanating from terminals in neighboring cells can be approximated to be constant (and equal to its average). What is the largest average number of terminals per cell ωA_c that can be admitted into the system, if we require the SIR constraint to be satisfied with 95% probability? What happens in the cases when the SIR requirement cannot be met?

9.2 We would like to compare the capacity of the uplink of a DS-CDMA-system with a system with orthogonal signalling and conventional cell planning. The available bandwidth allows for a processing gain of 128 in the DS-CDMA system and 128 channels in the conventional system. Assume that both cases require a SIR of at least γ_0 on the cell boundary. Assume further that we only need to consider the first tier of six interfering cells. In the conventional system ($K \geq 3$) we make the approximation that all the interference power from a cell comes from a point close to the access port (center of cell). In the DS-CDMA system ($K = 1$), this approximation would be too coarse. In addition the DS-CDMA system uses a power control scheme, striving to keep the received power constant at the intended receiver. The propagation loss grows with the fourth power of the distance. Estimate the largest average number of mobiles at 2% channel assignment rate.

9.3 In the DS-CDMA system in Example 9.1, the power control schemes exhibit a random control error. In fact, the received power from a mobile is not constant but given by

$$P_i = P_0 + \Delta P_i \ (\text{dB})$$

Where ΔP_i are independent (log-) normally distributed random variables with 0 dB (log-) mean and a (log-) standard deviation of $a = 2$ (dB). Let us for the sake of simplicity assume that the variations in the total interference power are small (and equal to its average).
a) Determine the capacity if the SIR requirement 7 dB is to be met with 95% probability?
b) How large fade margin does this correspond to?

9.4 Consider the uplink of a single-cell DS-CDMA system, where the processing gain is 20 dB. The number of users in the cell is Poisson distributed with the mean of 5. For a given instant, the probability that a randomly chosen user is active is 0.4. Target E_b/I_0 is 15 dB. The system conducts a perfect CRP power control (constant received power). Define that a user will be outage if the interference amount is 10 times greater than the background noise amount.

a) Determine the outage probability of a randomly chosen user.
b) Now, consider a multiple-cell system, where the other cell interference is 50% of the own cell interference. All the other parameters are the same as the above. Repeat question a).
c) Now, assume that the user activity probability becomes 0.2. Considering a single-cell system, determine the maximum E_b/I_0 that achieves the same outage probability as in a).

9.5 Consider the uplink of a multirate DS-CDMA system where the external (intercell) interference power can be assumed to be constant. Assume further that there are M users with spreading factors $N_1, N_2, \ldots N_M$. If the required SIR is γ_0, show that the optimum received power allocation for user i is proportional to

$$\frac{\gamma_0}{N_i + \gamma_0}$$

9.6 In the uplink of a DS-CDMA system the assumption that 50% of the interference at the base station is derived from mobiles outside the cell and that this level is roughly constant. In this exercise we will investigate when these conditions are valid. Assume a 4th-power of distance propagation law with 8 dB log-standard deviation and uniformly 2D-Poisson distributed active mobiles. Mobiles are power controlled (perfect CRP-power) by the base station with the lowest instantaneous path loss. Use RUNE to

a) Determine the ratio of (average) internal/external interference.
b) Plot the CDF of the interference power for the cases when the traffic load ωA_c is 1, 3 and 10 mobiles/cell.
c) In b normalize the interference power with traffic load (i.e., to display curves with the same average interference). What are your conclusions?

9.7 Consider an uplink of a cell in a DS-CDMA with the cell radius 1 km, where N users are randomly distributed. The users can make connections with four data rates (R, $R/2$, $R/4$, $R/8$). And, each data rate requires the same target $E_b/I_0 = 7$ dB. Assume $R = 9.6$ Kbps, and the spreading bandwidth of the system is 1.25 MHz. We will focus on a snapshot of the uplink situation. Assume a 4th-power of distance propagation law with 8 dB log-standard deviation and negligible receiver noise. Maximum power of each user is 1W; minimum is 0W. Base station receiver noise is 10^{-15}W.

a) Every user wants to transmit with the maximum rate R. Execute the DCPC power control algorithm until the iteration number reaches twenty (see Section 6.3.5). Draw a graph that shows the average data rate per user as a function of the iteration number. How does the graph change according to N?

b) Assume that at each iteration of DCPC, every user chooses its target data rate randomly from $(R, R/2, R/4, R/8)$. Now, repeat the job in a).

c) Compare a) and b) in terms of total throughput. What are your conclusions?

References

[1]   TIA/EIA Interim Standard-95: Mobile station-base station compatibility standard for dual-mode wideband spread spectrum cellular system, Telecommunications Industry Association, 1993.

[2]   Gilhousen, K. S., et al., "On the Capacity of a Cellular CDMA System," *IEEE Trans. Veh. Technol.*, Vol. VT-40, 1991, pp. 303–312.

[3]   Young, W. R., "Advance Mobile Phone Service: Introduction, Background and Objectives," *Bell Systems Technical Journal*, Vol. 58 (1), 1979, pp. 1–14.

[4]   CDMA Network Engineering Handbook, Qualcomm, Inc., San Diego, 1993.

[5]   Viterbi, A. M., and A. J. Viterbi, "Erlang Capacity of a Power Controlled CDMA System," *IEEE J. Sel. Areas Commun.*, Vol. SAC-11, 1993, pp. 892–900.

[6]   Berggren, F., and S. L. Kim, "Energy-Efficient Rate and Power Control in DS-CDMA Systems," submitted to *IEEE J. Sel. Areas. Commun*, 2000.

[7]   Dahlman, E., et al., "WCDMA—The Radio Interface for Future Mobile Multimedia Communications," *IEEE Trans. Veh. Technol.*, Vol. VT-47, 1998, pp. 1105–1118.

[8]   Viterbi, A. M., A. J. Viterbi, and E. Zehavi, "Performance of Power-Controlled Wideband Terrestrial Digital Communications," *IEEE Trans. Commun.*, Vol. COM-41, 1993, pp. 559–569.

[9]   Rulnick, J. M., and N. Bambos, "Power Control and Time Division: The CDMA Versus TDMA Question," *Proc. IEEE INFOCOM 1997*, 1997, pp. 634–641.

[10]   Hwang, S. H., et al., "Soft Handoff Algorithm with Variable Thresholds in CDMA Cellular Systems," *IEE Electronics Letters*, Vol. 33, 1997, pp. 1602–1603.

10

Resource Management in Packet Access Systems

10.1 Data Traffic and Performance Models

In the previous chapters, clever use of waveforms and power levels were studied in order to squeeze many simultaneous users into a given portion of the frequency spectrum. In most of the analysis, a single instant of time, a snapshot, where a set of terminals competes for the network resources, has been considered. This is indeed a reasonable model for conventional voice (conversational class) traffic, where the QoS requirements prescribe a channel with constant link quality and low, fixed delay. The classical information source model, emitting a "never ending" continuous flow of information symbols, is working nicely here with short to moderate periods of time (e.g., tens of seconds) (see Section 3.2). Not getting sufficient resources for such a session in a single snapshot may be sufficient to disrupt the service provided. If the attention is now turned to data traffic, in particular, the interactive traffic class (see Section 3.2), the picture is quite different. Here, the constant delay requirement, which has a fundamental impact on the design of radio resource management schemes, is dropped. A key feature of a data communication system is that the information stream is broken down into small, discrete units, messages, packets, or blocks that are required to arrive intact and in the correct order at the receiver but with somewhat different delays. Allowing slightly delayed data delivery opens up two possibilities:

1. Retransmissions of erroneous data (ARQ) can be used on the link allowing for lower link margins, tolerating more interference, and

259

still allowing for very low (residual) error rates. The reader is referred to [1, 2] for a brief recap of these techniques and their application to wireless data transmission.

2. The momentary unavailability of radio resources is no longer critical. This paves the way for adding time domain resource management (scheduling) to the RRM toolbox. Data messages can be queued for transmission at some later time when resources become available. This problem, known as the packet multiple access problem, is studied in some detail in Section 10.3.

Consider a modified communication model as outlined in Figure 10.1. Here M terminals, T_0 through T_{M-1}, are competing for the wireless resources. Again, as in Chapter 2, one of the terminals, say, T_0, is chosen to illustrate how a source data message is transmitted to the receiver. The figure also illustrates another key feature of a data communication system, the feedback channel/signal z, informing the transmitters whether or not a message transmission was successful. The terminals are capable of storing, or buffering, messages while waiting for the proper time to transmit the message.

As previously discussed in Section 3.2, modeling the sources $S_0 \ldots S_{M-1}$ for future wireless systems is difficult since there are many applications that have yet to be invented. There exists a span of traffic patterns ranging from an almost continuous stream of messages over a long time (i.e., streaming class audio/video or background class file transfers) to data-generating information in a more sporadic fashion (i.e., interactive class). In many situations a bursty source model, emitting messages at random intervals, may be more appropriate. A message would consist of a number of source symbols. Define such a simple model in which messages have constant size, N_m symbols. The message model could describe various data communication messages,

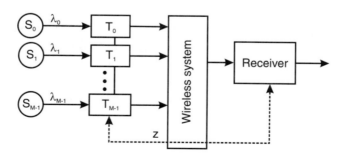

Figure 10.1 Message-oriented, multiuser communication model.

files, fax messages, and so forth. Now assume that messages are generated at the time instants a_1, a_2, a_3, ... where the sequence $\{a_i\}$ is a stochastic (point) process. Two types of message arrival models will be considered:

1. Interactive (Random) model: The arrival instants are derived from a Poisson process with intensity parameter Λ (messages/s). Remember that a Poisson process is characterized by the property that the number of messages that arrive during an interval of duration T is Poisson distributed with average ΛT. Furthermore, the Poisson process is memoryless, that is, no matter when we chose to observe the process, the time remaining to the next message arrival will be exponentially distributed with average $1/\Lambda$.

2. Streaming (Deterministic) model: The arrival instants are equally spaced in time $t_k = kt_0$. The arrival rate is Λ (messages/s).

Quite a few more detailed models have been proposed in the literature [3]. From the performance measures discussed in Section 3.2, the focus will here be on three types of measures. First, we will use the message delay (A). Again, denoting the arrival instant of the ith message at a terminal a_i and the instant when the same message is delivered by the receiver d_i, the delay of the ith message can be defined as:

$$D_i = d_i - a_i$$

The sequence of message delays will be a stochastic process. The delay may be caused by actual transmission delay (the time it takes to transmit the message over a wireless link with a given data rate), by retransmissions due to transmission errors, and finally by resource competition (waiting times in queues for available resources). If this process is stationary and has finite moments, that is, after sufficient time has passed

$$E[D_i] = E[D] = \overline{D}$$

independent of i, it is stated that the communication system is stable. Measuring delay and determining whether a system is stable or not are not trivial tasks. In an overloaded system (to high arrival rates), queues may build and the message delay sequence may have an ever-increasing average. In an ideal system, with infinite amounts of buffer storage, the moments of D will become infinite.

The second property that is considered is the quality of the data as it is received (B). This can be measured in different ways. One way is to use the residual error probability, which is defined as the probability that a given message at the receiver is in error at delivery, that is, the error was not detected (nor properly corrected). In most applications, this error probability has to be very low and is mostly used as a constraint. The use of error detection coding is a very powerful device to keep this type of quality impairment at bay at a very low cost in terms of added redundancy. More seriously, messages may be lost on their way. In practical systems with finite buffers, overloading will cause message dropping, that is, arriving messages may find the terminal buffer full and will be rejected. The normal procedure to model this type of system is to let D denote the conditional message delay, given that the message is not rejected, and let $\Lambda_{r,k}$ denote B, the message dropping rate of rejected messages for terminal k. This will be our second performance measure. Some applications do not allow any message dropping at all and will require the retransmission of all lost messages. The third performance measure considered is the *throughput* of the system (C). The individual throughput of messages, that is, the expected number of messages that are delivered by the receiver per time unit, is

$$S_k = \Lambda_k - \Lambda_{r,k}$$

This quantity is often also expressed in symbols/s or bits per second (multiplying S_k with the packet size). A stable system with infinite buffering is lossless, that is, the message dropping rate is zero and the throughput is simply equal to the arrival rate. The *system throughput*, finally, is then defined as the sum of the individual throughputs,

$$S = \sum_k S_k$$

Several formulations for the RRM problem in a message/packet-oriented environment are feasible depending on the designer perspective. Returning to the discussion on the role play between the consumer/user and the provider/operator, it is clear that measures A and B are consumer-oriented, describing the quality of service for an individual user. The system throughput (C) and in particular the maximum system throughput, which is a capacity measure, are definitely provider-oriented measures, since the revenues of the operators are to be derived form the number of messages (bits) they will transmit on behalf of the users. As in Section 3.1, the aim of this chapter is to study the tradeoff between these conflicting measures. The common design paradigm is again to maximize the system throughput, using the user performance measures as constraints. The resulting capacity may be expressed in terms of Kbps/km^2 in such a system.

The individual throughput is not that interesting per se since it is virtually independent of the message delay due to the buffering delays in many parts of the system. Alternative definitions of the individual throughput exist. One of the more popular measures used for file transfers is denoted as the link *throughput* and defined as

$$\tilde{\Lambda}(N) = E\left[\frac{N}{d_N - a_1}\right]$$

where N is the size of a file in messages and the denominator is the total time spent to transfer the file. In the definition above, it is assumed that message dropping can be neglected. The link throughput may also be expressed in bits per second. For large file sizes N, this performance measure approaches the individual throughput and the link throughput can therefore be used as an estimate of the latter. For example, it is this type of measure that is used in most Web browsers to indicate the current rate of data transfer. For small message sizes and large buffering delays, there may be a large discrepancy between them. Also, the *normalized link delay*

$$\tilde{D}(N) = \frac{1}{\tilde{\Lambda}(N)}$$

is sometime used.

Exercise 10.1

Show that $\tilde{\Lambda}_k$ approaches Λ_k for a lossless stable system with finite delay moments when the file size N grows.

10.2 Packet Multiple Access

In this section a closer look is taken at the third source of delay in wireless data communication systems, the resource-sharing conflicts. This is usually referred to as the *packet multiple access problem*. Initially, the treatment will be slightly simplified by investigating the uplink of a single-cell system, or a multipoint-to-point system. As will be seen, the uplink problem is indeed of a different character than the downlink (a point-to-multipoint system), since all resource management has to be performed by the terminals in a distributed manner. This is in particular a critical factor in systems with short messages and stringent demands on delay (response time in interactive system) where no time or system bandwidth can be wasted for coordinating the resource management.

In terms of the block diagram in Figure 10.1, the receiver will be the access port trying to receive the messages from all the different terminals. To keep the models simple, orthogonal waveforms will be used at this stage. In the following, TDMA waveforms are assumed, which is no further loss of generality. The system is lossless with an infinite buffer storage in the terminals. Further assume that the symbols are transmitted at a rate $1/T_b$, given that the whole channel bandwidth is available. The messages will thus have a duration of $T_m = N_m T_b$. To simplify the notation, all time measures will be normalized to this message duration T_m. In normalized time, the duration of a message transmission will thus be in one time unit. Now introduce the normalized intensity parameter, which for a lossless system is equal to the normalized individual throughput

$$s_k = \lambda_k = \Lambda_k T_m$$

expressing the message arrival rate in message per message duration. With this notation, the system throughput (again for the lossless system) can be expressed as

$$S = \lambda = \sum_k \lambda_k$$

The maximum throughput, S^* (or λ^*) is defined as the maximal value of S for which the moments of the expected queue length in the system remain finite. If we study the system for a long time, the number of messages entering the system and the number of successfully received messages must be equal in order to avoid excessive buildup of queues. Using the total available bandwidth for a single message, its transmission would require one time unit. The maximum number of packets that can depart from the system is thus one packet per time unit. Therefore, the following simple bound on the maximum throughput can be derived:

$$S^* < 1$$

Now, explore the possible RRM strategies that may be employed. For instance, split the available bandwidth/time slots into M_0 identical, orthogonal channels using one of the orthogonal multiplexing schemes discussed earlier. The symbol rate on each of these channels, $1/T_c$, is obviously a factor M_0 lower than if the entire bandwidth were used. The duration of a message in one of these subchannels is

$$T_m' = T_c N_m = M_0 T_b N_m = M_0 T_m$$

or, in normalized time

$$T = \frac{T_m'}{T_m} = M_0$$

The messages are generated by M sources with (normalized) message rates λ_0, λ_1, ..., λ_{M-1}. In the following, a communication system with symmetric traffic load is studied (i.e., a system where all $\lambda_i = \lambda/M$). The messages are fed to the M terminals, which by means of a resource sharing scheme, an access algorithm, will decide which (sub)channels to use and when. The transmitters often provide feedback information, Z, concerning the outcomes of previous transmission. The message communication system may be slotted, that is, message-synchronous where all messages have equal size and all transmissions start simultaneously at the beginning of given time frames. Systems that are nonslotted usually allow varying message size, and transmitters may start transmissions at any time.

In most cases $M > M_0$, which means that an arriving message may find all channels in use. Even if $M \leq M_0$, this could happen due to the fact that one of the message sources could momentarily generate more messages than the channel would be able to handle. A message arriving under such circumstances cannot be transmitted immediately. How that situation will be handled depends on the specific application. In computer communications, when some delay can be accepted, it is natural to let excess messages await their turn in queue. These messages will be transmitted when transmission capacity becomes available. In other applications when delay may be critical (i.e., telephone calls), the excess messages may be discarded, or blocked. The resources and the available channel bandwidth have to be shared between the arriving and waiting messages. At first glance, one may get the impression that optimizing this resource allocation with respect to message delay or with respect to maximum throughput would yield the same result. However, this is not the case, especially not for systems in which the total arrival rate, λ, is way below unity.

To illustrate different resource management strategies from a different perspective, take a simple example from economics. Study, for instance, the supply and demand for some simple commodity, say, soap. Two simple basic principles of providing soap could be imagined. In a planned economy approach one tries to estimate the demand beforehand and adjusts the production volumes for the next period according to this estimate. The production becomes very efficient and high volumes may be reached. If the demand is high and the demand estimates are accurate, very little resources are lost. However, if the actual demand does not match the estimates, there will either be surplus or a shortage of soap. In the market economy the production facilities receive constantly updated information about the demand for soap. This leads to a production that can be dynamically tuned to the actual demand. However, considerable resources are used to find this information in activities such as marketing (there will be 20 different brands in colorful wrapping), opinion polls, and so forth. The marketing system may not use all resources that efficiently, but will quickly react to changes in order to keep supply and demand in balance.

In communication systems we can think of two similar extreme strategies: fixed assignment (FA) schemes and dynamic assignment (DA) schemes. The model in Figure 10.1 can be used to illustrate these two principles. In a system with a fixed assignment, each user is tied to a channel, that is, $M_0 = M$. The user may use his or her channel at any time, but he or she will never be allowed to use the channel of a neighboring user, even if this channel is not in use. Since the bandwidth resource is allocated, once and for all, no portion of bandwidth has to be allocated for resource sharing purposes. Messages queue in every station and are transmitted totally independent of the other stations in the system. The message transmission process of each of the users can be modeled by what is in queuing theory denoted an M/D/1-queuing system. Such a system has Markovian (Poisson) arrivals, a deterministic (constant) transmission (service) time and one service unit (transmission channel). The system consists of M such independent queuing systems. This type of queuing system is well known in the literature [4]. It is fairly easy to derive the expected message delay in the system. If T denotes the normalized transmission delay in each channel, we can express the expected delay as[1]

---

1. This result is usually known as the Pollacek-Kinchin (PK) mean-value formula [4].

$$E[D_{FA}] = T + \frac{\lambda_i T^2}{2(1 - 2_i T)}$$

In the case with M identical stations and balanced traffic, $\lambda_i = \lambda / M$. Further, according to (7.15), $T = M$. Inserting these two results into the expression above yields

$$E[D_{FA}] = M + M + \frac{2M}{2(1 - x)} \qquad (10.1)$$

In a system with dynamic allocation, there are usually more users than channels, that is, $M_0 < M$. The channel are allocated to the active users, those that have a message to transmit. Available capacity is thus given to those users that are demanding it. However straightforward in principle, there are serious problems when applying this technique to radio systems. Radio systems are distributed and a station cannot definitely know which of the terminals may have a message to transmit. In most cases, some of the bandwidth of the system has to be assigned for the exchange of such demand information between the stations. This assignment overhead will be necessary to facilitate an orderly exchange of messages.

Figure 10.2 illustrates the performance of some typical FA and DA schemes. The graph shows the normalized expected delay $E[D]$ as function of the total arrival rate λ (i.e., the throughput). Note that systems with fixed assignment have a comparatively high delay at low arrival rates. In these cases, this system is practically empty, and only a few messages arrive from time to time. Since each station is allowed to use only its dedicated fraction of the bandwidth, the transmission time will be long. On the other hand, at high arrival rates, the entire bandwidth will be effectively utilized. The maximum throughput for this system λ_{F^*} is high. The system using dynamic allocation will have almost all the bandwidth at its disposal and yields a very low delay at low traffic loads. When the system is empty, an arriving message can be transmitted at once at a considerably higher rate than in the FA system. On the other hand, when the traffic load is increased, more resource sharing transmissions are required and the maximum throughput λ_{D^*} is considerably lower than in fixed assignment. Figure 10.2 clearly illustrates how the dynamic channel assignment algorithm attempts to achieve a low delay at moderate loads at the expense of a lower maximum throughput.

The amount of bandwidth required for the assignment overhead depends on how rapid the changes in demand actually are. In systems with short, bursty messages, the relative overhead is larger than in a system with

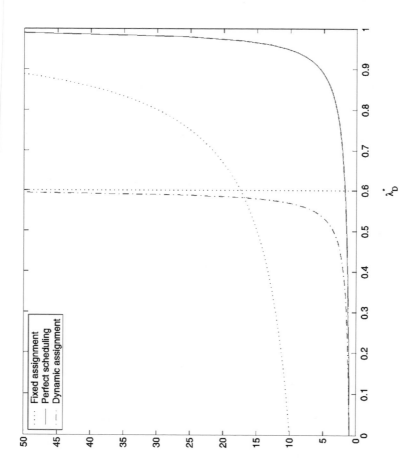

Figure 10.2 Normalized expected delay, $E[D]$, as functions of message arrival rate (throughput) for systems with fixed assignment (FA) and dynamic assignment (DA) in comparison with perfect dynamic assignment (PS) ($M = 10$).

long messages, since the allocation process has to be repeated often. In the sample system in Figure 10.2, it could be concluded that the assignment overhead is about 50% of the required bandwidth. The third graph in Figure 10.2 shows the performance of a hypothetical system with no assignment overhead. This is termed a *perfect scheduling* (PS) scheme. Such a system would be realizable only if all stations could instantaneously know the queuing situation in all other stations. In this way, a single queue could be arranged and messages could be transmitted in order without loss or additional overhead. This hypothetical system is used to provide a lower bound on the expected message delay, since messages in the PS system can be regarded as placed in a single queue ($M_0 = M = 1$) for all messages. Using the result in (10.1) and applying $T_k = M = 1$ and $\lambda_i = \lambda$ will yield

$$E[D_{PS}] = 1 + \frac{\lambda}{2(1 - \lambda)} \tag{10.2}$$

that is,

$$E[D_{PS}] = \frac{E[D_{FA}]}{M}$$

The D_{PS} delay as a function of λ is shown in Figure 10.2. The asymptote of the delay, the maximum throughput, is the same as for the fixed assignment case (i.e., $\lambda_{PS}^* = 1$).

In almost all modern communication systems, one tries to emulate a dynamic assignment schemes, that is, striving for some type of adaptation of the resource allocation to the present demand. Most of these practical schemes fall into the category of reservation schemes. In these schemes, a part of the bandwidth is used to collect information about the communication demands of the terminals and the rest of the bandwidth is used for perfectly scheduled transmissions of the messages. The latter can be done on separate channels or on dedicated timeslots. The fraction of the bandwidth used to collect sufficient information is equivalent to the allocation overhead mentioned. Such a scheme works excellently when messages are long or for long voice sessions. The amount of bandwidth spent to make the reservation can in these cases be negligible. On the other hand, when messages are few and short, most of the bandwidth is spent in the reservation process, yielding very low throughputs. For these types of applications it is often favorable to design systems such that $M \gg M_0$. In fact, the simplest system design is achieved when $M_0 = 1$. The distributed schemes that are studied in the following let every terminal decide when to access the channel based only on their own observations. There is an obvious risk of a conflict when several terminals decide to access the channel simultaneously due to their imperfect knowledge about the current state in other terminals.

A specific model describing this situation, known as the *collision channel with feedback*, will now be studied. Compared to the damage done by collisions, noise will be rather harmless and initially the impact of the latter will be neglected. If a single station transmits a message, the message will reach the receiver successfully. If, however, two or more stations transmit at the same time, conflict occurs and none of the messages are assumed to be received correctly. After each transmission attempt, each station will receive information about the outcome through some error-free feedback mechanism. An example of such a mechanism is the transmission (or lack) of acknowledgment messages. Assuming that only very short acknowledgment messages are used, the bandwidth required for this feedback channel can be neglected. The feedback, in Figure 10.1 modeled as the variable Z, is

0 No transmission;
1 Successful transmission;
2 Unsuccessful transmission attempt (collision).

This feedback information plays a crucial role in a dynamic resource assignment scheme. In fact, the sequence of feedback signal is the only available information for the stations to make their transmission. Such a decision-making algorithm is denoted an access algorithm. Access schemes for this particular channel models are usually called Random Time Division Multiple Access (RTDMA) algorithms. Finding the algorithm that minimizes message delay or maximizes throughput is still an unsolved problem. In the following some simple schemes of practical importance are studied.

The ALOHA Algorithm

The first algorithm in the class of RTDMA schemes, later coined the "ALOHA algorithm," was proposed in the late 1960s by researchers at the University of Hawaii [5]. The algorithm was devised for use in a VHF-radio system to connect remote terminals on the many islands with a central computer site. In the original scheme the stations were not synchronized. The feedback information was used in the simplest way:

1. If there is a message—transmit it!
2. In the case of a successful transmission—remove the message from the queue and return to 1.

3. In the case of a collision—wait a random time interval and go to 1.

It is of paramount importance that the stations do not wait identical time intervals after a collision, since this would with certainty cause a new collision. The method to avoid this in the ALOHA algorithm is for all stations to perform, independently of each other, a random experiment to decide when to transmit the next time. The scheme is quite practical in the case when M is large.

A simple modification of the basic ALOHA scheme is to confine stations to start their transmissions at the beginning of well-defined slots with duration equal to the of message transmission time. Let us analyze the performance of such a slotted ALOHA system. For this system, the algorithm may be simplified even further. Let each station that has a packet to transmit "toss a coin" with success probability p at the beginning of each slot. If the toss is successful, the station will transmit the message in the queue. This variety of the ALOHA scheme has roughly the same properties as the original scheme.

Making an exact analysis of this scheme is beyond the scope of this text. Instead, some further simplifications and rough approximations will be made. The results will still give a fairly accurate description of the system performance. Study the model in Figure 10.3 describing the flows of messages in one of the many (M) stations in the system. As previously assumed, the message arrives (from the left) according to a Poisson process with rate $\lambda_i = \lambda / M$. As soon as a message arrives, it is transmitted in the next time slot. Depending on the other transmission attempts in the system, the transmission may or may not be successful. If it is successful the message will arrive at the receiver and leave the system to the right. A collided message will be stored in the delay buffer. Message k will remain there X_k (a random

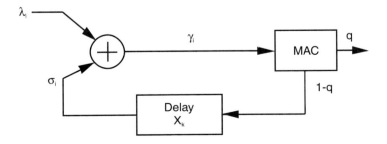

Figure 10.3 Simplified transmitter model in slotted ALOHA system.

number of) time slots. The stochastic variables X_k are assumed to be independent and identically distributed. Upon leaving the delay buffers these messages are transmitted immediately. Let us make the approximation that the delayed messages leave the delay buffer according to a Poisson process with rate σ_i. Further, one may note that this process is independent of the arrival process. A consequence of the approximation is that the message arrivals will be memoryless, independent of whether the message is new or if it has been waiting for retransmission, or backlogged. According to the law of large numbers, this approximation is accurate if

$$E[X_k] = E[X] = \xi \gg 1$$

where the simpler notation X is used to denote the stationary process X_k. Now the compound process of retransmitted and new messages will also form a Poisson process with intensity γ_i given by

$$\gamma_i = \lambda_i + \sigma_i \qquad (10.3)$$

Denote by q the probability that station i makes a successful transmission in a given slot. If the system is in a stable equilibrium, the average number of messages arriving to the system has to be equal to the average number of departing messages. Otherwise, the system will either contain an ever-increasing number of packets or become empty. This condition can be expressed as

$$q\gamma_i = \lambda_i \qquad (10.4)$$

Turning to the probability p, the condition for a successful transmission is that no other station is transmitting in the same time slot. Since all stations are identical, the probability that a station will transmit will be the same for all stations. Let us by p denote the transmission probability. Thus,

$$q = (1 - p)^{M-1} \approx (1 - p)^M \qquad (10.5)$$

If the messages arriving to be transmitted arrive according to a Poisson process with rate γ_i, the probability p can be expressed as

$$p = \Pr\{\text{at least one arrival in the slot}\}$$
$$= 1 - \Pr\{\text{no arrival in the slot}\} \qquad (10.6)$$
$$= 1 - e^{-\gamma_i}$$

Inserting this result into (10.5) yields

$$q \approx (1 - p)^M = e^{-M\gamma_i} = e^{-\gamma} \tag{10.7}$$

where $\gamma = M\gamma_i$. Now, combine (10.4) and (10.7) to get

$$\lambda_i = q\gamma_i = \gamma_i e^{-\gamma} \tag{10.8}$$

Summing (10.8) over all M identical stations yields

$$\lambda = \sum_{i=1}^{M} \lambda_i = M\gamma_i e^{-\gamma} = \gamma e^{-\gamma} \tag{10.9}$$

Equation (10.9) describes a relation between the total message arrival rate, λ, and the total rate of transmission attempts, γ [Figure 10.4(a)]. In the literature this quantity is often referred to as the offered load of the system. An interesting observation is that the RHS of (10.9) cannot take arbitrarily large values. In fact, λ has a maximal value given by

$$\lambda \leq \max \gamma e^{-\gamma} = e^{-1} \tag{10.10}$$

The RHS is the maximal throughput $\lambda^* = e^{-1}$.

We may now also estimate the message delay. First, condition the delay on the number of retransmission attempts required for a successful transmission. This number of attempts is denoted N. Every retransmission attempt requires one slot for the actual transmission, plus a stochastic delay X. The normalized message delay is derived as

$$E[D] = \sum_{n=0}^{\infty} E[D|N = n]P[N = n] = \sum_{n=0}^{\infty} (1 + nE[X])P[N = n]$$

$$= \sum_{n=0}^{\infty} P[N = n] + E[X] \sum_{n=0}^{\infty} nP[N = n] = 1 + \xi E[N] \tag{10.11}$$

Since a transmission is successful with probability q, the number of retransmissions N is geometrically distributed. Here, the approximation that successive transmission attempts are independent has been made. The latter is again a reasonable assumption if $\xi \gg 1$. The expectation computes as

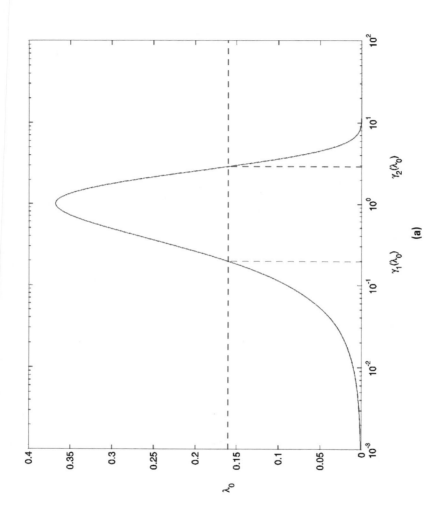

Figure 10.4 (a) Throughput as function of transmission attempt rate (offered load) for slotted ALOHA algorithm (7.27); (b) Normalized message delay in slotted ALOHA as function of total message arrival rate λ (10.12).

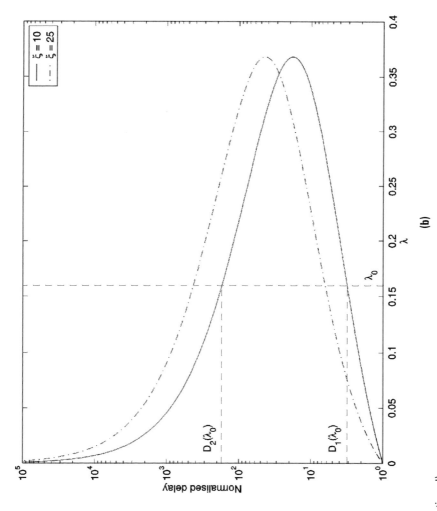

Figure 10.4 (continued).

$$E[N] = \frac{1}{q} - 1$$

Inserted into (10.11) yields

$$E[D] = 1 + 5\left(\frac{1}{9} - 1\right) = 1 + \xi(e^{\gamma} - 1) \qquad (10.12)$$

Together with (10.9), this constitutes a parametric description of the delay as a function of the arrival rate, using the parameter γ. To be able to derive a numerical value for the delay a function of λ, we need to start by determining γ. Solving for γ in (10.9) provides an unpleasant surprise: In general, there is not only one but two solutions. Denote these by $\gamma_1(\lambda)$ and $\gamma_2(\lambda)$. We denote the delays that can be computed from (10.12) by $D_1(\lambda)$ and $D_2(\lambda)$. These relations are illustrated in Figure 10.4(b) where the two delays are plotted as one continuous curve. The curves corresponding to the two solutions meet for $\lambda^* = e^{-1}$, corresponding to the delay $E[D] = 1 + \xi e \approx \xi e$.

For every message arrival rate, there are two possible delays. One of these delays, $D_1(\lambda)$, is comparatively low, whereas the other $D_2(\lambda)$ in general is considerably higher. There are obviously two states of equilibrium. The state of the system corresponding to $D_1(\lambda)$ corresponds to a low channel activity with almost all transmissions being successful. In the other, high delay state, there are a large number of backlogged messages in the system, causing frequent retransmission. The attempt rate γ is high, but the success probability q is low. Due to the normal stochastic variations in the input message stream, the system will move between these two equilibrium states. It is possible to show that the probability of ending up in the high delay state decreases with increasing ξ. A large ξ thus has a stabilizing effect on the system, but will unfortunately also cause a large average delay, as in (10.12). Further, one can show that for systems with finite M, the system can always be made completely stable (a single equilibrium state). However, this is again achieved at the expense of an impractical high delay. For a more detailed analysis of the stability properties of ALOHA systems, the reader is referred to [6].

The calculation of the throughput and delay is illustrated briefly by means of a simple example:

Example 10.1

In a fleet management system, a number of mobile stations transmit requests to a central base station. The request messages are of constant size, 200 bytes (1,600 bits). There are 60 mobiles, each making a request according to a Poisson process with a rate of one request per minute. The modulation scheme offers a maximal total data rate of 9,600 bps. When designing the system, there is a choice between a fixed assignment scheme using 60 slots per frame, and a RTDMA system using a slotted ALOHA algorithm. The average retransmission delay in the latter is assumed to be $\xi = 25$. (Use the low delay equilibrium in your calculations.)

a) Which of the systems provides the lowest expected message delay?

b) At which request rate will the systems have the same average message delay?

Solution:

We first compute the message duration and the normalized arrival rate:

$$T_m = NT_b = 1600/9600 = 1/6\text{s}$$
$$\lambda = M\Lambda_i T_m = 60 \cdot 1/60 \cdot 1/6 = 1/6$$

a) ALOHA: Solve (numerically) for γ in $\lambda = \gamma e^{-\gamma}$ (10.9)

$$\lambda = 1/6 \qquad \gamma \approx 0.204$$

(10.12) yields

$$E[D_A] = 1 + \xi(e\gamma - 1) \approx 6.7 \text{ slots} \approx 1 \text{ sec}$$

TDMA: (10.1) yields

$$E[D_T] = M\left(1 + \frac{\lambda}{20 - x}\right) \approx 66 \text{ time slots} \approx 10 \text{ sec}$$

Answer: The ALOHA system has the lowest average delay.

b) Solve the following equation (numerically)

$$D_A(\lambda_0) = D_T(\lambda_0)$$

$$\lambda_0 \approx e^{-1} \approx 0.367 \approx 2.2 \text{ requests/(minute } \cdot \text{ mobile)}$$

λ_0 is very close to the maximum throughput and the ALOHA system is thus very close to becoming unstable at this arrival rate. The TDMA system is stable for all $\lambda < 1$, that is, for arrival rates less than 6 requests/(minute $\cdot$ mobile).

In a wireless network, the simple assumption that a packet is lost if more than one terminal is transmitting and that no packet is lost if only one terminal is transmitting, has obvious flaws. Certainly a packet may be lost due to unfavorable propagation conditions (e.g., fading), even though a single terminal is transmitting. In this respect, the assumption is optimistic. However, if several terminals are transmitting, one of them may be received at a considerably higher power level than all the others such that the signal-to-interference ratio for the reception of a packet from this particular terminal is still feasible. This effect is usually referred to as power capture. The simple assumption of completely destructive collisions is in this respect somewhat pessimistic. To take these two phenomena into account, the model has to be elaborated somewhat. Let the received signal-to-interference ratio for receiving a packet from terminal i be

$$\Gamma_i = \frac{P_{rx,i}}{\displaystyle\sum_{\substack{k \in K \\ k \neq i}} P_{rx,i} + N}$$

where $P_{rx,i}$ is the received power from the terminal and K is the set of the colliding packet. Assume that the packet from terminal i is *captured* and correctly received whenever $\Gamma_i > \beta$, where β is the SIR threshold or "capture ratio." Making a further simplification, it is assumed that the probability that more than two terminals are involved in a collision is negligible. This is a reasonable assumption under light-to-moderate load conditions. Thus, simply write the condition for a successfully receive packet from terminal 1 as:

$$\Gamma_1 = \frac{P_{rx,1}}{P_{rx,2} + N} > \beta \qquad (10.13)$$

Assume that without a loss of generality that terminal 1 and 2 are colliding. Note that a packet will be successfully received if

$$\Gamma_2 = \frac{P_{rx,2}}{P_{rx,1} + N} > \beta \qquad (10.14)$$

in which case a packet from terminal 2 is captured by the receiver. If we use the geometry in Figure 10.5 with a circular cell area with radius R_{max}, neglect the noise N, and use a simple r^α propagation model, the *capture probability* may be computed (i.e., the probability that one of the packets is successfully received), as

$$P_{cap} = 2\Pr\left[\frac{P_1}{P_2} \geq \gamma_0\right] = 2\Pr\left[\left(\frac{R_2}{R_1}\right)^\alpha \geq \gamma_0\right]$$

$$= 2\Pr[R_1 \leq R_2(\gamma_0)^{-1/\alpha}]$$

$$= 2\int_0^{R_{max}} \Pr[R_1 \geq R_2(\gamma_0)^{-1/\alpha} \mid R_2 = r_2] f(r_2) dr_2$$

If the terminals are uniformly distributed over the circular cell area

$$P_{cap} = 2\int_0^{R_{max}} \Pr[R_1 \leq r_2(\beta)^{-1/\alpha}] f(r_2) dr_2$$

$$\qquad (10.15)$$

$$= 2\int_0^{R_{max}} \frac{r_2^2(\beta)^{-2/\alpha}}{R_{max}^2} \frac{2r_2}{R_{max}^2} dr_2 = 2\left(\frac{1}{2}(\beta)^{-2/\alpha}\right) = \frac{1}{\beta^{2/\alpha}}$$

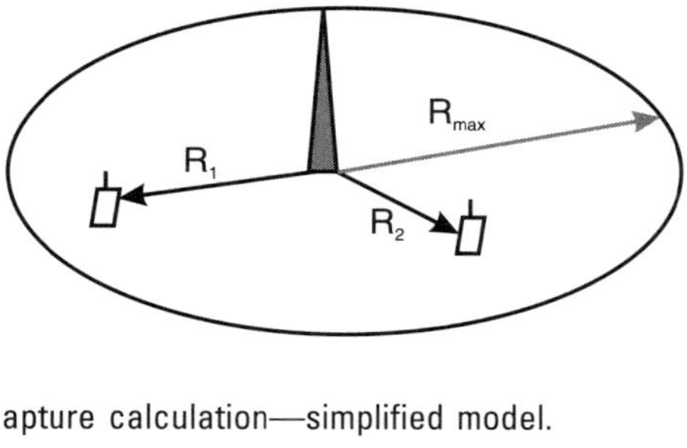

Figure 10.5 Packet capture calculation—simplified model.

Modifying (10.6–10.8) yields

$$\lambda = q\gamma = \gamma \Pr[\text{successful transmission}]$$

$$= \gamma(e^{-\gamma} + \gamma P_{cap} e^{-\gamma}) = \gamma e^{-\gamma}\left(1 + \frac{\gamma}{\beta^{2/\alpha}}\right) \qquad (10.16)$$

As the capture ratio increases, the expression approaches (10.9). In fact, $\beta = \infty$ (for negligible N) corresponds to the case where a packet can be received only if there is no other transmission going on at the same time. $\beta = 1$ corresponds to perfect capture, where the strongest packet always will be properly received.

An interesting generalization is $\beta < 1$ where more than one packet can be received successfully (corresponding for instance to some spread spectrum modulation scheme). Figure 10.6 shows a plot of (10.16). It can be seen that the maximum throughput could almost be doubled, compared to the results with the previous model, if efficient modulation and detection schemes can be used.

As these simple results indicate, the variations in power due to the near-far effect work really to advantage of the system, which stands in bright contrast for instance to the situation in CDMA uplink systems. In fact, the more the signal level fluctuates, the better the performance. Certainly, large variations will occasionally drop the signal level below the noise floor causing a noise-related packet loss. However, it has been shown in [7] that there is still a substantial net gain even if the noise is taken into account. In the problem section we will also illustrate this. Further, we have made the simplifying assumption that only two terminals are involved in a collision. In [8] a more general approach is taken and the probability that a packet is captured in a k-fold collision is computed. A last complication, which our model does not capture, is the fact that in a real wireless network, with a finite traffic load, the offered traffic will be dependent of the location in the cell. This is due to the fact that terminals close to the access port will have a higher probability of being successful, and will therefore have its packets delivered more quickly and present a very limited amount of retransmitted packets. Terminals close to the perimeter of the cell will, however, need to perform a lot of retransmission since they tend to "loose" in most collisions. The effect of this is treated in more detail in [9].

The observation that differences in the received power level can be advantageous has interesting implications regarding the power control strategy that should be employed. In the previous sections it was demonstrated

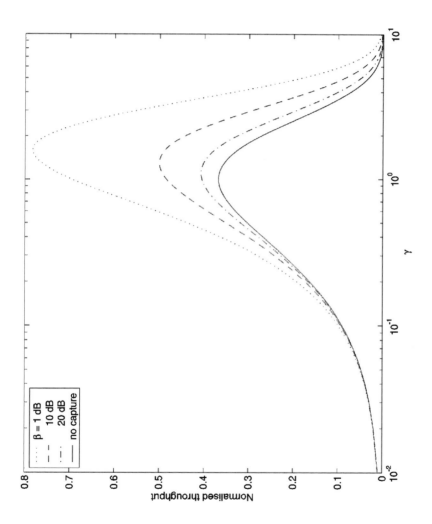

Figure 10.6 Throughput versus offered traffic for different capture ratio in the single-cell slotted ALOHA system.

that equal balanced SIR or equal received power strategies were the successful ones. In this context, however, they utterly fail. This can be easily understood since they almost guarantee that all packets in a collision will be lost. Instead, power control schemes that even exaggerate the differences have been proposed and shown to provide large performance gain [10]. The transmitter power dynamic range mainly limits the performance of such a scheme. The reader is referred to the problem section for further illustration of this principle.

Carrier Sense Multiple Access (CSMA)

This class of access algorithms is popular in asynchronous (nonslotted) radio systems with low propagation delays. Each of the stations is equipped with a receiver in order to monitor if any transmissions are in progress in the radio channel. The stations simply measure the signal level to detect transmission, that is, carrier sensing. If there is a carrier present (transmission in progress), the station defers its transmission attempt to a later time. In other respects the CSMA algorithm works in the same way as the ALOHA algorithm: If a collision occurs, the retransmission is delayed by a random time interval. One usually distinguishes between two types of CSMA schemes: persistent and nonpersistent schemes. The nonpersistent scheme will treat the detection of an ongoing transmission the same way as a collision and the station waits a random amount of time before sensing again. In contrast, a persistent algorithm will upon the detection of a transmission wait until the end of that transmission and then transmit its own message.

An important parameter in a CSMA system is the detection delay of the system. This is the time span between when a station decides to start a transmission until the instant when all stations become aware of this transmission (and thus prevent a collision). This delay includes the carrier measurement delay, the switch time from reception to transmission, and the propagation delay. If the detection delay is large compared to the message size, the carrier sensing information will be of little use since it is only able to describe the state of the channel some time ago. The capability to prevent collisions deteriorates and two stations may both sense that the channel is empty and start their respective transmissions. However, if the detection delay approaches zero, almost all collisions can be avoided.

Performance-wise, the CSMA algorithm constitutes a significant improvement over the ALOHA scheme, provided that the detection delay can be kept low. For high detection delays the sensing information becomes useless and the performance approaches nonslotted ALOHA. CSMA systems exhibit the same bistable behavior as ALOHA, and the stability is affected by the average retransmission delay in much the same way. The performances of some CSMA schemes and nonslotted ALOHA are compared in Figure 10.7. In the graph the relative detection delay is $a = 0.05$, that is, the delay constitutes 5% of a message duration.

CSMA schemes are very popular in VHF/UHF ground packet radio systems where all stations need to communicate with all other stations. Another well-known application is wireless local area networks, which will be discussed in the following section.

Conflict Resolution Algorithms

Both the ALOHA and the CSMA algorithms have very primitive methods of resolving a conflict. The solution used in these schemes is simply to postpone the transmission indefinitely (a random amount of time) while hoping for the best. A number of more systematic schemes to resolve conflicts have been proposed, beginning in 1979 when Capertanakis presented his tree-search algorithms [11]. This scheme has been further refined and is now held to be among the best-known algorithms (with respect to maximum throughput), including the Tsibakov-Michailov's part-and-try or stack algorithm [12]. The principle used in these schemes is simple: Stations are numbered with unique integers 0, 1, 2, . . . , $M < q^K$ (where K should be chosen as small as possible). The numbering is performed in the base q (i.e., every station identity is a K multidigit number where the digits are chosen in the range 0, 1, . . . , $q - 1$). The simpler systems have two modes—one free access mode where new messages may enter as they arrive, and a conflict resolution mode, where only messages involved in a collision may participate. As soon as a conflict occurs, the system enters the conflict resolution mode and remains in this mode until the conflict has been resolved. Then it returns to the free access mode.

In the conflict resolution mode, a systematic conflict resolution algorithm (CRA) is employed. In the first time slot after the collision, only stations having a zero as the final digit in their station IDs are allowed to transmit. If another collision occurs, the station with final digits equal to 00 is enabled to transmit. However, if the next slot is empty, or if it contains a successful transmission, we proceed with stations with IDs ending with a 1. For every collision the number of ending digits are increased to solve a partial conflict. The method can be described as searching a tree according to Figure 10.8. Every node in this tree corresponds to a time slot. The root is the time slot corresponding to the free access mode. At every collision a

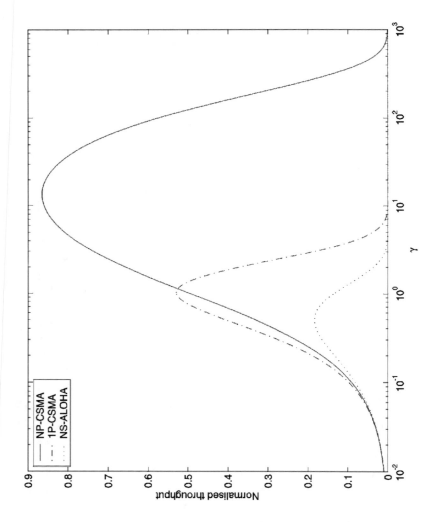

Figure 10.7 Throughput for some CSMA algorithms, $a = 0.01$ [4].

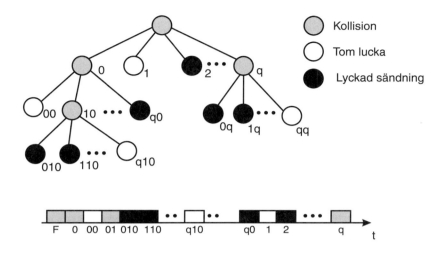

Figure 10.8 Tree search CRA. Tree and corresponding slot allocations.

new set of q branches is created. The tree may be searched according to the depth first scheme, but other search strategies will also achieve the same maximum throughput. Empty slots and successful transmissions form the leaves in the tree. When we have reached leaves in all branches, the conflict is resolved and the CRA stops.

The maximum number of nodes that have to be searched is given by

$$N_{\max} = \sum_{i=0}^{\overline{K}} 9^i = \frac{9^{\overline{k}+1} - 1}{9 - 9} \tag{10.17}$$

The choice of q is determined by the delay and the maximum throughput that is required that is to be achieved. In general, it could be stated, that for $q = 2, 3$, good delay properties are achieved for moderate loads, whereas for $q > 4$, a high maximum throughput is sought. By selecting a fairly large q and skipping the free access state, we can go directly into the CRA without waiting for a collision. Such a system corresponds to a multichannel system where $M_0 = q$. The stations use a subchannel according to their final digit in their ID. On each channel we will perform the same tree search scheme as previously. By skipping the free mode we can gradually move towards a fixed assignment scheme. In fact, this could be done adaptively, matching the current traffic load. Note that the special case $q = M$ corresponds to a system with pure fixed assignment.

Reservations Schemes

In systems with long messages and messages of strongly varying length, the so-called reservation schemes have become increasingly popular. These schemes are often used in single channel situations ($M_0 = 1$). Since it is very costly to lose a long message in a collision, one tries to avoid collisions in these systems. For this purpose, stations compete for channel resources, using short reservation packets. A reservation system operates in two modes: a reservation phase and a data phase (see Figure 10.9). During the reservation phase, some type of RTDMA algorithm (e.g., ALOHA or CRA) is used to transmit the reservation packets. The reservation packets are as small as possible and contain only information regarding the station identity, the number of pending messages, and their size. When the reservation phase is over, the result is announced by a central controller or deduced by the stations themselves in a distributed system. The result is the reservation of a number of slots in the data phase that now follows. In the data phase, only messages that have received reservations in the previous phase are allowed to transmit. The data phase will thus be collision-free. The duration of the data phase may well be variable and is in general adapted to the number of requested data slots. In this case the data phase ends when all messages that were able to obtain reservations have been transmitted, very efficient in achieving a high maximal throughput. The result is that the message delay will increase. To keep the delay under control, the length of the data phase may be limited.

In systems in which the reservation and data phases have fixed lengths, one may discuss performance in terms of reservation capacity and data capacity. The reservation capacity is the number of reservations the system can handle per time unit, whereas the data capacity is the amount of messages or bits that may be transmitted per time unit. It is clear that the smallest of these quantities will limit the performance of such a system. In a well-designed system, there is a balance between the reservation capacity and the data capacity. To increase the reservation capacity, we need to increase the number of reservation slots, which will clearly be at the expense of the number of slots assigned for data. Since the reservation messages are that

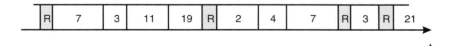

Figure 10.9 A typical slot layout in a reservation system.

short, there is usually no problem in providing ample reservation capacity, at least if the message size is large.

Making a general analysis of the performance of a reservation system is somewhat complicated and beyond the scope of this text. The reader is again referred to [6] for a more thorough treatment. Here, only an approximate method is used, where the system is studied under light load and heavy load conditions. A constant message size and a variable data phase duration are assumed. During the data phase, it is assumed that all stations that are succeeding to get reservations have managed to empty their queues during the data phase. Assume that there will be B messages transmitted during a data phase, where B is a stochastic variable. Assume further that a CRA is used during the reservation phase and that a reservation slot has a duration δ. Let N_{res} be the number of slots used in the reservation phase. This latter number will also be a stochastic variable.

The maximum throughput of this system may be derived from the high load approximation. High load means that queues will build in all stations and that the data phase will be long. In equilibrium the average number of messages that enter the system has to be equal to the average number of messages that leave the system. Since the data transmission phase is collision-free, the maximum throughput is only the fraction of time spent in the data transmission mode. The result is

$$\lambda \geq \frac{E[B]}{E[B] + \delta E[N_{res}]} \qquad (7.32)$$

Note that for a finite number of stations, N_{res} may always be finite. As the data phase grows in length with higher λ, the RHS expression approaches 1. The maximum throughput is thus unity ($\lambda^* = 1$). The fraction of time spent in the reservation phase vanishes.

If the system is studied at low loads, the data phase will be very short. The majority of data phases will, in fact, have zero duration. The typical delay for a single message arriving under these conditions consists of the duration of a reservation phase and the time spent to transmit the actual message. The reservation will take only one reservation slot of duration δ. The expected delay becomes

$$D(\lambda \approx 0) \approx 1 + \delta$$

10.3 Some Packet Access Applications

In the following section the packet access features of two current wireless standards are briefly outlined to illustrate some of the principles described in the previous sections.

10.3.1 WCDMA—Packet Access

The UMTS/3GPP WCDMA system is a cellular system of conventional structure, where terminals communicate with the access ports (base stations) of the fixed network. The general transmission formats and characteristics of the system have been outlined in Chapter 9. For packet access (uplink), two access principles are available: a packet by packet random access mode using a random access channel (RACH), and a reservation mode where a dedicated channel (DCH) is allocated for the duration of the data transfer. Figure 10.10 illustrates the operation in the two modes. In the "random access" mode, the user messages are fragmented in to radio blocks and transmitted in the 10 ms "radio frames" of the RACH. The encoding allows for a variable data rate to handle varying transmission conditions. The same channel (codes and frames) is used by several users in a random access fashion and collisions are a real threat.

As in the ALOHA schemes with capture discussed in the previous section, the near-far property of the system will cause packets be received at quite varying power levels. As we have seen, this increases the probability of capture if all users were packet users using the same access technique. In the WCDMA system, however, several different radio bearers have to coexist. Sometimes the received power level will be too low, compared to the interference from other services, and the achieved SIR will be below the threshold. Since packet transmissions are short, there is no way to control the transmitter power to an adequate level during the actual packet transmission. The

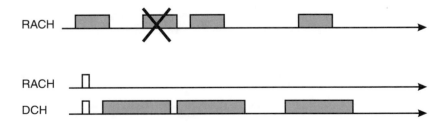

Figure 10.10 WCDMA packet access modes.

alternative to use higher transmitter power is detrimental to other coexisting services.

In order to handle the power control problem and increase the success probability, a preamble power ramping scheme has been devised. The scheme is illustrated in Figure 10.11. Instead of transmitting the message immediately, the terminal transmits a sequence of short preamble bursts with increasing power. When such a preamble is properly received by the base station, an access indicator is transmitted by the base station and the terminal proceeds with transmitting the packet with the power level determined by the ramping procedure. In order to limit the number of collisions, a number of measures are used. First, the base stations controls the transmission probability p that is transmitted on the downlink at regular intervals. Further, the network can determine that only terminals of certain access service classes should be allowed to transmit. In addition, the terminal may use different preamble spreading codes and may also select the exact time slot for transmission. The main advantage of RACH mode for data transfer is that it requires no setup and is ideally suited for short packets with low delay requirements. The drawback is that each packet has to carry explicit address information. In addition to the ramping scheme, this creates a rather large overhead and makes the achievable relative throughput moderate.

The alternative, which is particularly attractive for longer messages, is to set up a dedicated channel for the terminal in question. This is done with a sequence of handshake messages on the RACH and the DCH. Once the setup is made and the DCH becomes available to the terminal, all the RRM attributes of the WCDM system (see Chapter 9) are available, that is, fast and slow power control, link rate adaptation, and so forth. As with all reservation schemes, once the reservation is made, throughput is high since no explicit addressing is needed and very little other overhead is required. The price paid is, of course, the setup time, which can be in the order of several tens of milliseconds.

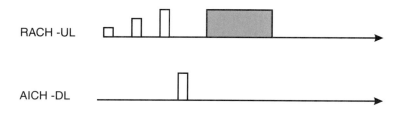

Figure 10.11 RACH power ramping technique.

10.3.2 IEEE 802.11 Wireless Local Area Network (WLAN) Access

In contrast to the WCDMA system described previously, an IEEE 802.11 WLAN is a distributed system, in which there is no clear master-slave relationship between terminals and the access port. Basically, a WLAN is an ad-hoc network allowing peer-to-peer communication between all terminals and access ports (the latter ones may provide connections to some wired infrastructure, but do not control the access procedure in the same way as in a cellular system). The access scheme for the 802.11 has similarities to the 802.3 Ethernet wired line standard and uses a protocol known as carrier-sense, multiple access, collision avoidance (CSMA/CA). This protocol avoids collisions instead of detecting a collision like the algorithm used in 802.3, because it is difficult to detect collisions in a wireless network since terminals are not able to receive signals while they are transmitting due to excessive dynamic range requirements. A clear algorithm is used to determine if the channel is clear by measuring the RF energy at the antenna and determining the strength of the received signal. If the received signal strength is below a specified threshold, the channel is declared clear and the terminal is given the clear channel status for data transmission. If the RF energy is above the threshold, data transmissions are deferred in accordance with the protocol rules. The 802.11 CSMA/CA protocol allows for options that can minimize collisions by using request to send (RTS), clear-to-send (CTS), data, and acknowledge (ACK) transmission frames in a sequential fashion (see Figure 10.12). Communications is established when one of the wireless nodes sends a short message RTS frame. The RTS frame includes the destination and the length of the message. The message duration is known as the network allocation vector (NAV). The NAV alerts all others in the medium to back

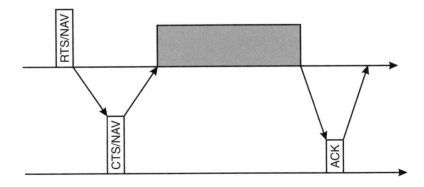

Figure 10.12 IEEE 802.11 CSMA/CA scheme.

off for the duration of the transmission. The receiving station issues a CTS frame, which echoes the sender's address and the NAV. If the CTS frame is not received, it is assumed that a collision occurred and the RTS process starts over. After the data frame is received, an ACK frame is sent back verifying a successful data transmission. A common limitation with wireless LAN systems is the hidden node problem. This can disrupt 40% or more of the communications in a highly loaded LAN environment. It occurs when there is a terminal in a service set that cannot detect the transmission of another terminal to detect that the wireless channel is busy. In Figure 10.13, terminals A and B are capable of communicating. However, an obstruction prevents terminal C from receiving signals from terminal A and it cannot determine if the channel is busy. Therefore, both terminals A and C could try to transmit at the same time to terminal B. The use of RTS, CTS, data, and ACK sequences helps to prevent the disruptions caused by this problem. The reader is referred to [13] for a more detailed description.

Problems

10.1 In a data communication system for taxis, a radio channel providing a bit rate of 4,800 bps is used. The messages consist of only 50 characters (8 bits/char). A slotted ALOHA access system is used.
   a) Determine an upper bound on the number of taxis that can be served by this system if every taxi can be assumed to generate one message (according to a Poisson process) every minute.
   b) Estimate the expected packet delay in a system with 100 taxis if the average retransmission delay is three seconds!

10.2 In a slotted ALOHA system with N terminals, the terminals control their power such that the received power takes one out of K predetermined values, $P_1, P_2, \ldots, P_K$. Determine the optimal values P_i and estimate the maximum throughput for
   a) $K = 2$
   b) $K = N$

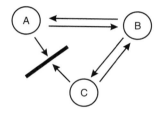

Figure 10.13 Hidden terminals in CSMA schemes.

10.3 In a slotted ALOHA system collided packets are all lost and single transmitted packets are assumed to be lost and retransmitted with probability P_e due to noise.
   a) Determine an expression for the maximum throughput.
   b) Derive and plot the delay as a function of the throughput (corresponding to the favorable equilibrium) for P_e = 0, 0.05, 0.1, 0.25. Use ξ = 20 in your calculations.

10.4 An synchronous RTDMA system with four terminals uses a conflict resolution algorithm (CRA) with a binary tree (q = 2). Assume that all terminals, independently of each other, have a message to transmit at the beginning of a conflict with probability p.
   a) What is the maximum number of slots that are required to resolve a conflict?
   b) Determine the expected duration of the conflict resolution process in time slots as a function of p.
   c) Determine the expected duration of the conflict resolution process if the first time slots are consequently skipped (M_0 = 2). For which values of p is the system better than the original system?

10.5 In a single-cell, slotted ALOHA system, terminals are communicating with their base station with packets of constant size. The receiver noise can be neglected, the path loss is r^4-distance dependent with no additional fading. Packets are successfully received if the signal-to-interference ratio is larger than 10 dB. The probability of correct packet reception when more than two packets are colliding can be neglected. The system uses the following transmitter power control regime:

$$P_{tx} = P_0 Z$$

where Z is a $U(\epsilon, 1)$ distributed random variable ($0 < \epsilon < 1$), drawn for every transmission. Compare the maximum throughput of this system with a system using no power control ($P_{tx} = P_0$) and a system with constant received power control.

10.6 In a single-cell slotted ALOHA system, terminals are communicating with their base stations with packets of constant size. Packets are successfully received if the signal-to-interference ratio is larger than 10 dB. The receiver noise can be neglected, and the path loss is log-normally distributed with an 8-dB log-standard deviation and an

r^4-distance dependent median loss. In order to provide maximum fairness in the system, the terminals control their power in order to compensate for the path loss. Due to the sporadic nature of the Poisson packet traffic, the terminals move considerable distances between subsequent transmissions. Therefore, terminals are not able to measure the path loss perfectly, but manage only to compensate for the median loss (which can be assumed to be perfectly estimated).

a) Neglecting the probability that more than two packets are involved in a collision, determine the maximum throughput of the system.

b) Assuming perfect path-loss compensation (i.e., including the log-normal variations), what is the maximum throughput now?

10.7 In the uplink of a mobile packet radio system, a large number of terminals access a central station using a slotted ALOHA multiaccess scheme. Since the terminals are at different distances from the central receiver, they are received at different signal levels. If the ratio between the signal level of the strongest received packet and the other packets is large enough, the strongest packet may still be received even if there is a collision ("capture"). Assume that the signal level decays as the fourth power of the distance and that the noise can be neglected within the circular design coverage area. Packets arrive according to a Poisson process, with a uniform traffic load distribution over all the terminals. Assume that no terminal can be closer to the base station than a fraction d of the cell radius.

a) Determine the distribution of the received power from a terminal.

b) Determine/estimate the maximum throughput of the uplink of the system provided that the probability of correct packet reception when more than two packets are colliding can be neglected (the Goodman/Saleh model). Assume an SIR threshold for the correct reception of 5, 10, and 15 dB. Compare results for $d = 0$ and 0.1.

c) Check your results with a simulation where any number of colliding packets is considered. Use the Namislo model and as a first step estimate (by simulation) the probabilities $q_i = \Pr[\text{success in collision of multiplicity } i]$ for the same SIR requirements as i, b). Plot also this intermediate result and compare it with the results in [8] for $d = 0$ and 0.1.

10.8 Using the (infinite population) Goodman/Saleh model [9], estimate and plot the packet delay for a single cell system measured as the number of transmission attempts per successful transmission for terminals at a given radius r. Assume SIR requirements 5, 10, and 15 dB. Plot results for $S = 0.1$, 0.3, and 0.5.

10.9 In the uplink of a mobile packet radio system, a large number of terminals access a central station using a slotted ALOHA multiaccess scheme. Since the terminals are at different distances from the central receiver, they are received at different signal levels. If the ratio between the signal level of the strongest received packet and the other packets is large enough, the strongest packet may be received even if there is a collision (capture). Two systems using binary modulation, the same number of information bits per packet and the same total bandwidth, are being compared: a conventional system requiring an SIR of 10 dB for the correct reception of a packet, and a direct sequence–modulated system with a processing gain of 10. Assume that the signal level decays as the fourth power of the distance and that the noise can be neglected within the circular design coverage area. Packets arrive according to a Poisson process, with a uniform traffic load distribution over all the terminals. Neglecting collisions of more than two packets, estimate the maximum throughput in packets per time unit for both systems.

References

[1]   Ahlin, L., and J. Zander, "Principles of Wireless Communication," *Studentlitteratur*, 1997.

[2]   Lin, S., and J. Costello, *Error Control Coding: Fundamentals and Applications*, Englewood Cliffs, NJ: Prentice-Hall, 1983.

[3]   Paxson, V., and S. Floyd, "Wide Area Traffic: The Failure of Poisson Modeling," *IEEE/ACM Trans Networking*, Vol. 3, No. 3, June 1995.

[4]   Kleinrock, L., *Queueing Systems, Part I: Theory*, New York: John Wiley & Sons, 1976.

[5]   Abramsson, N., "ALOHA Packet Broadcasting,"

[6]   Rom, R., and M. Sidi, *Multiple Access Protocols*, New York: Springer-Verlag, 1990.

[7]   Arnbak and Blitterswijk, "Capacity of Slotted ALOHA in Rayleigh Fading Channels," *IEEE Trans Sel. Areas Comm.*, SAC-5, February 1987.

[8]   Namislo, "Analysis of Mobile Radio Slotted ALOHA Networks," *IEEE Trans. Veh. Tech.*, VT-33, No. 3, August 1984.

[9]   Goodman and Saleh, "The Near/Far Effect in Local ALOHA Radio Communications," *IEEE Trans. Veh. Tech.*, VT-36, No. 1, February 1987.

[10]   Metzner, J. J., "On Improving Utilization in ALOHA Networks," *IEEE Trans. Comm.*, COM-24, April 1976.

[11]   Capertanakis, J. J., "Improvements in Block Retransmission Schemes," *IEEE Trans. Comm.*, COM-27, February 1979, pp. 524–532.

[12]   Tsibakov, B.S., and V.A. Michailov, "Random Multiple Access: Part-and-Try Algorithm," *Probl. Inform. Transmission*, Vol. 16, No. 4, Oct.–Dec. 1980.

[13]   Wireless LAN Alliance (WLANA), www.wlana.com.

11

Cell-Planning

System engineers spend a considerable amount of time configuring a radio network in terms of location of radio access points (base stations), frequency assignment, power levels, types, and heights of antennas. This procedure is known as *cell-planning*. User traffic distribution and the radio propagation conditions of the service area have to be considered as accurate (reliable) as possible during the cell-planning procedure. This information cannot be a perfect level, however, because traffic and propagation conditions vary considerably in both time and spatial domains. On the other hand, fine-tuning of these parameters, for example, locations and powers (coverage) of base stations will greatly influence the overall performance of the cellular system.

In this chapter, some of key concepts of cell-planning will be covered. Since cell-planning itself is a complex problem, it is hard to provide any perfect solution for the cell-planning problem. Rather, this chapter will provide a few insights into the problem by explaining some issues.

11.1 The Cellular Concept

Over the years, engineers set various objectives for designing high-capacity cellular radio systems. When the first commercial cellular system, AMPS (Advanced Mobile Phone Service) was designed [1], those engineering objectives included:

- Large subscriber capacity;
- Efficient use of spectrum;
- Adaptability to traffic density.

Although these criteria are more than 20 years old, they are still applicable to today's cellular radio systems. In those early days, the need to meet the above goals given a limited number of channels was the strong driving force behind the evolution of the cellular concept. This cellular concept was verbalized in 1947 by D. H. Ring of Bell Laboratories, even though there was no published document available to us [1]. As has been identified in Chapter 4, there are two major aspects in the cellular concept: frequency reuse and cell splitting. In fact, frequency reuse was initially employed in broadcasting and other radio services. Employing frequency reuse in a mobile communication service required exploitation of a shrunken geographical area (cell). Through frequency reuse, a cellular radio system can handle a large number of simultaneous calls that exceeds the total number of allocated frequencies.

As was described in (4.14) in Chapter 4, there are several factors that determine the cellular system's capacity. The cellular system is basically *scaleable* so that its capacity depends particularly on the total number of cells over the service area. It is sufficient to split a larger cell into smaller ones to cope with the increased user demand without increase in the spectrum allocation. Cell splitting helps to meet the objective of being adaptable to the spatial density of demand for channels. Lower-demand areas can be served by larger cells (macro), whereas higher-demand areas are served by smaller cells (micro). More detailed discussion about the system of mixed cells will be given Section 11.3.

11.2 Cell-Planning Based on Regular Hexagon Geometry

Even though the cellular concept is simple, cell-planning in a real situation is not that easy. Because of propagation variation, it is not possible to precisely define a coverage area for a given cell. As was discussed in Section 4.2, this drives the system engineer to visualize cells as regular hexagons, where each base station is assumed to transmit with equally strong power. Furthermore, with the help of geometry on regular hexagons, system engineers can simplify the cell-planning problem (as was described in Section 4.2) summarized by

$$\frac{R}{D} = \sqrt{3K}$$

where K is given by

$$K = (i + j)^2 - ij, \; i, j = 0, 1, 2, \ldots$$

It maximizes the channel reuse distance while cochannel interference (channel quality) is acceptable. However, this approach obviously has some limitations. Among those, the most critical aspect is that it cannot handle traffic unbalance among the cells.

In Section 4.2, it is assumed that the traffic distribution of each cell is identically distributed, giving the same traffic density to every cell. For instance, consider the cell-planning in Figure 11.1. Since the same number of channels is allocated to each group of cochannel cells, it cannot cope with the traffic unbalance among the cochannel cells. This results in unused capacity in some cells and shortage of channels in the other cells.

However, while keeping the simplicity of the regular hexagonal cell-planning, a new frequency assignment that can cope with the traffic nonuniformity may be developed [2]. For example, with a given cluster size $K = 7$, two different frequency assignments can be constructed. This is done by determining the nearest cochannel cells from the current cell:

- Moving j cells in the clockwise direction after moving i cells along any chain of hexagons (Figure 11.1 (a));
- Moving j cells in the counter-clockwise direction after moving i cells along any chain of hexagons (Figure 11.2 (a)).

In the cases such as $K = 3, 4, 9, 12, \ldots$ it makes no difference between clockwise and counter-clockwise patterns. In that case, it is desirable to adjust

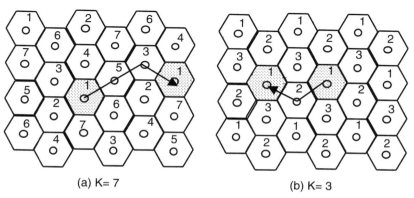

(a) K= 7   (b) K= 3

Figure 11.1 Cell-planning based on the regular hexagons.

it to the smallest value such that $K = (i + j)^2 - ij$ has two different solution pairs at least, while satisfying the reuse distance constraint.

Now overlap Figure 11.1(a) onto Figure 11.2(a). Then, the result will be the one in Figure 11.2(b), where two channel groups, one from the clockwise pattern and the other from counter-clockwise pattern, are allocated to each cell. The total available channels are divided into 14 groups, and channels are assigned to each cell according to Figure 11.2(b). This method will open the possibility of assigning the different numbers of channels to each cochannel cell.

An interesting question is how to assign channels to each of the 14 groups in a way that the total number of channels is minimized while covering the traffic demands of all cells. This problem can be formalized as follows. In Figure 11.2, assume that each cell i has channel requirement η_i. Furthermore, assume that $C_1(i)$ and $C_2(i)$ denote the channel group indexes that are assigned to cell i from the clockwise pattern and the counter-clockwise pattern, respectively, that is, $C_1(i) \in \{1, 2, \ldots 7\}$ and $C_2(i) \in \{8, 9, \ldots 14\}$. Denote the number of channels that will be allocated to the channel group j by $D(j)$. Then the goal is to decide in a way that:

$$\min \sum_{j=1}^{14} D(j)$$

subject to

$$D(C_1(i)) + D(C_2(i)) \geq \eta_i, \forall\ i,$$
$$D(j) \in \{0, 1, 2, \ldots\}, j = 1, 2, \ldots 14.$$

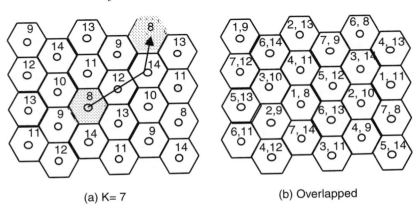

(a) K= 7   (b) Overlapped

Figure 11.2 Regular hexagon cell-planning—traffic distribution consideration.

The problem looks like a combinatorial optimization problem at first glance. However, it is proved that such an assignment can be found by solving a linear programming problem [3].

11.3 Hierarchical Cell Structure

In a coverage area where two or more sizes of cell exist simultaneously, special care must be taken to guarantee the minimum reuse distance between the cochannel cells D/R. In Figure 11.3, the cell **a** makes a gradual transition (cell splitting) from microcell to three microcells, **a1**, **a2**, and **a3**. In order to keep the minimum reuse distance, the channels assigned to cell **a** may be divided into three groups, in which each group corresponds to a microcell, respectively.

From this solution, however, there is no gain in terms of capacity increase. In the case when the same channel can be reused between the cells **a1** and **a3**, then the channels in cell **a** can be divided into two groups, leading to a capacity enhancement. However, invoking the hierarchical cell structure (HCS) [4] leads to another solution that could result in much higher trunking gain. This concept recognizes that, when multiple cell sizes coexist, the cellular pattern is viewed as the superposition of a microcell pattern on top of a macrocell pattern. Implementing this concept may require that the total channels are divided into two groups, one for microcell layer and the other for the macrocell layer.

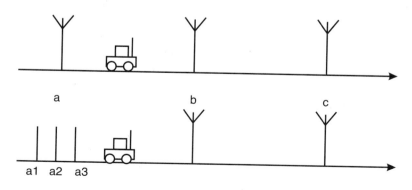

Figure 11.3 Cell-splitting and mixed cell sizes.

Example 11.1 Simple capacity calculation of HCS (Channel separation between layers)

Consider the cellular system of Figure 11.3. Assume that the total 12 channels are equally assigned to three cells, **a**, **b**, and **c** (four channels per cell). Now, compare this system's capacity with that of an HCS, where nine microcells, (**a1**,**a2**,**a3**), (**b1**,**b2**,**b3**), and (**c1**,**c2**,**c3**) are overlaid on top of three macrocell, **a**, **b** and **c**, respectively. Assume that in the micro layer, the same frequency channels can be (re-) used if cells are separated by two cells. For instance, **a1**, **b1**, and **c1** can be cochannel cells.

Solution:

Divide the twelve channels into two groups: one for the macro layer and the other for the micro layer. Then, to each macrocell, two channels can be equally assigned. In the micro layer, two channels (6/3) can be assigned to each micro cell. Therefore in total, eight channels are available to the coverage area of cell **a** with HCS, whereas four channels are originally assigned to that area.

In HCS, the cell layers may have different characteristics in terms of supported services, coverage, etc. An example of HCS is a system, where macrocells are used for *coverage* and microcells are used for *capacity*. In this system, the operator may choose to connect different users to different cell layers depending on user behavior and service requirements.

11.3.1 Channel Separation in HCS

A macrocell/microcell overlay system may be considered with a common frequency band in both layers, for instance when DS-CDMA is used in both layers. However, due to fast power control, mobiles connected to macrocells cause a lot of interference to users in microcells. This will be more severe to the microcell users near the macrocell border. A particularly difficult situation arises when a fast moving terminal connected to a distant macrocell enters a microcell. The unwanted radiation of the high power terminal into the microcell band may cause severe difficulties to the micro cell BS receivers that tune to the low power microcell terminals. This forces terminals currently communicating with a macrocell to be connected to a microcell whenever they get close to it. Otherwise, they will ruin the connections for the microcell users. However with this handoff scheme, since the terminal moves very fast, it results in undesirable frequent handoffs and increasing control traffic. Therefore microcells can only be sparsely distributed among macrocells, which cannot be regarded as a true example of HCS. On the contrary, with the channel separation, a mobile close to a microcell can be connected to a macrocell. Due to this ability of a free choice between the cell layers, one can keep a mobile with high speed connected to macrocells, in order to prevent frequent handoffs. Moreover, some mobiles can be given restricted access physically to just one cell layer. For instance, some services may be supported by only one cell layer.

An analysis on the interference between layers is given in [5], when micro and macro layers shares frequency spectrum with:

- System I: DS-CDMA in macrocells and TDMA in microcells;
- System II: TDMA in macrocells and DS-CDMA in microcells;
- System III: DS-CDMA in both macrocells and microcells.

Even though the System III is superior in terms of capacity to the other two systems, in general, microcells suffer much interference from macrocells due to power control. The microcells capacity is decreased, particularly when

- Number of macrocell users increases;
- The microcell is close to a macrocell's boundary.

In ETSI WCDMA system, interfrequency (hard) handover between the layers is supported by the slotted downlink mode [6]. When in the downlink slotted mode, the base station sends data with a faster speed, transmitting a 10-ms data frame in less than 10 ms. In this situation, an idle period is created, during which no data is transmitted by the base station. This period is then used for the terminal to tune to the other frequencies and measure signal strength.

Although the channel separation between cell layers is assumed, there still remains a key problem; because of the large dynamic range of received signals, the suppression of inter-band interference may not be sufficient to provide adequate signal-to-interference ratios at all times. A DS-CDMA system actually causes a lot of out-of-band interference. In some cases, the interference may be very severe, especially to microcells. Therefore a guard band between the cell layers will be required. In order to be spectrally efficient, it is desirable to keep the separation between the frequency bands at a minimum. An analytical expression and simulations have been developed on this problem for UMTS [7].

11.3.2 Velocity-Based (Interlayer) Handoff

Microcell/macrocell overlay system proposals often assign mobile users to a particular layer according to their speed. Fast users are generally encouraged to join macrocells, and slow users typically join microcells. Users are then instructed to move between the microcell and macrocell layers based on their speed. This jointly reduces the number of intralayer handoffs and increases the total system capacity. However, since mobile speed is varied constantly depending on spatial conditions and user characteristics, special care should be taken to prevent frequent interlayer handoffs. For this purpose, there has been some study on velocity-based (interlayer) handoff algorithms [8, 9]. In general, however, it is a complex problem to design an efficient interlayer handoff algorithm in the sense that the total number of handoffs (interlayer and intralayer) is minimized.

The velocity of a mobile is measured approximately by simply gathering the time spent in a cell by the mobile. A more accurate estimation of the vehicle velocity is possible if the received Doppler frequency is known. There is a useful relationship between the branch switching rate of a diversity receiver and its Doppler frequency, which permits the estimation of vehicle speed without any significant hardware changes [10]. There have been a few methods for velocity estimation proposed (e.g., level crossing rates, zero crossing rates, and a covariance method), and their performance differs depending on the propagation environment [11].

11.4 Automatic Cell-Planning

In cell-planning, engineers should consider the constraints such as operator's budget, availability of frequency bandwidth, availability of base station location sites, availability of wireline networks, propagation-based measures, required QoS (i.e., signal to interference ratios, new call blocking, handoff failure, forced termination probability, and carried traffic).

System engineers make large efforts in cell-planning, either in the initial stage of the system or in keeping up with the increase in traffic demand. This process, sometimes performed manually, is based on prior experience and intuition and has to go through several iterations before achieving satisfactory performance. Conventionally, regular reuse patterns are utilized and then manually optimized by using propagation and traffic statistics. However, substantial improvements in the planning process may be possible by designing algorithms to perform the optimization process automatically. Initially, when the demand for wireless services was low, the manual planning approach could be employed with a reasonable amount of confidence. However, the explosive growth in traffic has led to an increase in the density of cell sites. This results in greater network complexity, making it extremely difficult to manually plan cell sites for optimum performance. Therefore manual planning methods are being replaced by automatic cell planning techniques [12, 13]. The automatic planning tools take advantage of powerful optimization algorithms that enable near-optimum solutions through the evaluation of a large number of potential cell sites in a relatively short time.

References

[1]   MacDonald, V. H., "The Cellular Concept," *Bell Systems Technical Journal*, Vol. 58 (1), 1979, pp. 15–37.

[2]   Zhang, M., and T. S. Yum , "The nonuniform compact pattern allocation algorithm for cellular mobile systems," *IEEE Trans. Veh. Technol.*, Vol. VT-40, pp. 387–391, 1991.

[3]   Kim S. L., and S. Kim, "A Two-Phase Algorithm for Frequency Assignment in Cellular Mobile Systems," *IEEE Trans. Veh. Technol.*, Vol. VT-43 (3), 1994, pp. 542–548.

[4]   I, C.-L., L. J. Greenstein, and R. D. Gitlin, "A micro/macro cellular architecture for low- and high-mobility wireless users," *IEEE J. Sel. Areas Commun.*, Vol. SAC-11, 1993, pp. 885–891.

[5]   Wu, J. S., J. K. Chung, and M.-T. Sze, "Analysis of uplink and downlink capacities for two-tier cellular system," *IEE Proc.-Commun.*, Vol. 144, 1977, pp. 405–411.

[6]   Dahlman, E., et al., "WCDMA-The radio interface for future mobile multimedia communications," *IEEE Trans. Veh. Technol.*, Vol. VT-47, 1998, pp. 1105–1118.

[7]   Karlsson, R., and J. Zander, "Hierarchical cell structures for FRAMES wideband multiple access," ACTS Mobile Summit, Granada, Spain, Nov. 1996.

[8]   Ivanov, K., and G. Spring, "Mobile speed sensitive handover in a mixed cell environment," *Proc. IEEE VT 1995*, Chicago, IL, 1996, pp. 892–896.

[9]   Benveniste, M., "Cell selection in two-tier microcellular/macrocellular systems," *Proc. IEEE Globecom 1995*, Singapore, 1995.

[10] Kawabata, K., T. Nakamura, and E. Fukuda, "Estimating velocity using diversity reception," *Proc. IEEE VTC 1994*, Stockholm, Sweden, 1994, pp. 371–374.

[11] Austin, M. D., and G. L. Stuber, "Velocity adaptive handoff algorithms for microcellular systems," *IEEE Trans. Veh. Technol.*, Vol. VT-43, 1994, pp. 549–561.

[12] Sherali, H. D., C. M. Pendyala, and T. S. Rappaport, "Optimal location of transmitters for micro-cellular radio communication system design," *IEEE J. Sel. Areas. Commun*, Vol. JSAC-14, 1996, pp. 662–673.

[13] Hao, Q., et al., "A low-cost cellular mobile communication system: A hierarchical optimization network resource planing approach," *IEEE J. Sel. Areas. Commun*, Vol. JSAC-15, 1997, pp. 1315–1325.

12

Some Fundamentals of Wireless Infrastructure Economics

12.1 Telecommunication Infrastructures

Although this book has mainly covered the efficient utilization of the spectrum resources, the scope should be widened somewhat, taking some other aspects into account. As was pointed out in the introduction (see Figure 1.2) there is a close relationship with other resources as well. The wireless infrastructure (networks, switches, routers, and access ports) and the terminals have been covered. Here, the tradeoff between adding more access ports (i.e., a more expensive infrastructure and increased capacity) and a higher QoS provided to the users will be reviewed in some more detail. Comparisons of diverse quantities such as frequency allocations, infrastructure and real estate, user equipment, and user satisfaction are naturally made in economic terms.

The telecommunications industry is a global one with many actors. Whereas the supply chain of content, services, and transport services have been outlined in Figure 3.3, Figure 12.1 gives a rough description of the actors on the scene and their interrelations. In the early days of telephony (late 19th century), telecommunication companies (telcos) emerged that provided a complete service to the customer. They provided the wiring, the handsets, and the (manual) switching service. Every telephone company provided its own equipment since there was no standardization. Local competition between competing operators in the same neighborhood was fierce and no regulation was imposed. Situations where several competing telcos

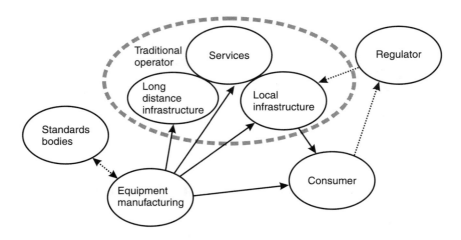

Figure 12.1 Actors on the telecommunication scene.

wired the same neighborhood led in many cases to a cost crisis, in particular where the wiring costs were high and the customers few. Not one of the competing telcos achieved sufficient scale advantages to become profitable, which in turn resulted in one of two possible outcomes: either the companies merged (acquired one another), or the local law makers granted one of the operators an exclusive license to operate a telephony system. In both cases a monopoly was formed where a single telco was operating all telephony services in some local area. In many European countries, this local monopoly evolved into a nationwide one, where all of the telecommunication services in the whole country were offered by one (in most cases state-owned) company. It is often debated whether telephony services constitute what economists call a natural monopoly. The latter is a technical term roughly describing those situations where a single company is capable of providing all the services demanded by the consumers at a lower cost than two or more companies (due to economies of scale). In the real world things are more complex. The regulator, which is aiming at protecting the public interests, has other objectives besides ensuring that the service is provided at a minimal cost. Most regulation has also a public welfare objective of providing telecommunication service in a fair and affordable way, such that everyone, wherever he may live, has access to telephony at a price that he can afford (universal access). The consequence is that in some places the service is made available at a low price where the marginal cost of providing it is much higher. These situations are usually called a market failure, since under normal market conditions no operator would provide the service (at those prices) in the area in question. The most important tool available to the lawmaker has been the licensing of monopoly rights.

The sale of monopoly rights has a long history as a way of raising money for the government or monarch of the state. This, of course, has nothing to do with protecting the public interest, but is still an important element in countries needing development finance. The licensing regime may produce such benefits as

- Creating a natural monopoly where costs are lower;
- Limiting the interference from unrestricted use of the electromagnetic spectrum;
- Imposing public service obligations which would not emerge from market processes;
- Imposing quality and standards.

Licensing can take place in many ways. Extremes include auctioning (grant the license to the highest bidder) and beauty contests where licensees are chosen based on the trust the regulating body puts in the licensees' ability to fulfill the license agreement. The national monopolies (natural or imposed by the licensing process) were intact in most countries for most of the 20th century. Regulation and telecommunication policymaking is a delicate matter. The intention is to apply controls and incentives to make the monopolistic operator behave as if operating in a fully competitive market. The operators adjust themselves to the rules and the result may not always be what the regulator originally intended. The monopolist is required to provide all the services (and almost everywhere), both the profitable ones, as well as the unprofitable ones, at equitable (and in many cases tightly regulated) prices. With the evolution of telecommunication technology, severe discrepancies between the cost of providing the service and the prices charged became obvious. The most obvious example is long distance telephony. In the early days, providing the wiring for long distance telephony was expensive and the cost was virtually proportional to the distance. With today's technology, a single physical cable and all of the required switching equipment is shared by thousands of simultaneous telephone and data connections such that the cost is almost independent of the distance. For the local access network, on the other hand, not much has changed. The individual consumer still has to be reached by a twisted pair of wires for exclusive use. With the evolution of technology, advanced network elements (switches, routers, and so forth) have become less and less expensive, whereas the cost of wiring (in particular for local access) is dominated by labor cost, that is, increasing over time. This can be illustrated with the following simple numerical example:

Example 12.1

The two cities, A and B provide telephony to all its households, both local and long distance connections between both cities (see Figure 12.2). The cities have 100,000 households each, and traffic to the outside world can be neglected. Assume that the network cost is dominated by the cost of providing the wiring and that this cost is proportional to the length of the cables. Estimate the relation between the cost of the access network (last 100m to the consumer) and the intercity network.

Solution:

At a cost of c currency units per meter, the total cost of the wiring in the access networks is

$$C_A = 2 \cdot 100,000 \cdot c \cdot 100 = 2 \cdot 10^7 c$$

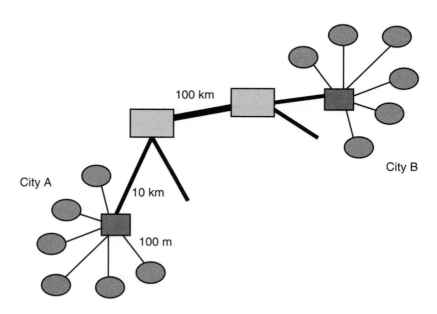

Figure 12.2 Access versus long-distance network comparison.

The intercity network wiring cost is

$$C_I = c \cdot 100 \cdot 10^3 = 10^5 c$$

We have that

$$\frac{C_A}{C_I} = \frac{2 \cdot 10^7 c}{10^5 c} = 200$$

Thus, the cost of the access network is 200 times higher than the intercity link. The intercity link is a 1,000 times more expensive than the individual access link for each household but its cost can be said to be shared by 200,000 households.

Even though example 12.1 is somewhat exaggerated, it still illustrates a few key points. The fact that nowadays, long distance communication is cheap to provide and local communication is expensive, whereas current prices indicated the opposite. This made long distance telephony the first target of emerging new telcos as soon as the licensing regulation was relaxed in the 1980s. As the wave of deregulation and competition swept over the world, most of the monopoly telcos business areas, have become the business opportunity of the incumbent operators. The local infrastructure, which the consumer has learned to view as a public utility, virtually free of charge, has for natural reasons had a very low priority among the new actors and very few new investments have been made in this area. Existing (former) monopoly operators for many years controlled an extensive access infrastructure with huge sunk investment costs, effectively preventing market entry for new actors at any significant scale. One exception here is the rapid spread of cable television, which today reaches more than 50% of the households in most western world countries.

12.2 Wireless Access Systems

The advent of cellular technology at the beginning of the 1980s marked a significant change in the telecommunication infrastructure business. The preceding decades witnessed several breakthroughs in telecommunication technology such as optical fiber communication, satellite communication, and the digital computer. These advances mainly improved the economics of long distance communications but did very little to change the conditions in the access networks, where the bulk of the total investment lies. Wireless technology, and in particular, cellular telephony made a radical difference since it had the potential of not only offering market entry opportunities for telecom access with significantly lower infrastructure investments, but also provided additional service quality in terms of mobility and roaming. This section probes a little more deeply into the economic driving forces behind the unprecedented success-story that is mobile (cellular) telephony.

In order to do this, start by reviewing (4.14). For the sake of simplicity it is assumed that a fixed modulation scheme and a given quality of service (blocking) is provided. The system capacity can in this case be written as

$$\omega \approx c'(\alpha)\frac{C\varpi_\eta}{\gamma_0^{2/\alpha}A_c} = c''\frac{C}{A_c} = c''\frac{C}{A_{tot}}N_{AP}\ (\text{Erlangs/km}^2) \qquad (12.1)$$

where A_{tot} is the area to be covered by the total system and N_{AP} is the number of (uniformly distributed) access points. What can be noted in (12.1) is that the number of users that can be served is directly proportional to the number of access points. The cost of providing the infrastructure to meet the service requirements will include the cost of providing the base stations (including base station sites, antennas, towers, and so forth) and the fixed wired communications network connecting the base stations. The following analysis will not consider the cost of the terminals. In the simplest cost model it is assumed that the cost of (building and maintaining) the infrastructure is simply a linear function of to the number of base stations, that is,

$$\textit{Cost model I: } C_{infra} = c_1 + c_2 N_{AP} \qquad (12.2)$$

In many cases the bandwidth used by the system will not be free (see licensing, bandwidth auctions). In this case it is reasonable to assume that there will be an additional cost proportional to the bandwidth

$$\textit{Cost model II: } C_{sys} = c_1 + c_2 N_B + c_3 W_{sys} \qquad (12.3)$$

of the system that is dedicated to a single user (except the wireless terminal itself). All parts of the infrastructure are shared.

Combining (12.2) with (12.1) the relative cost per user can be computed:

$$\frac{C_{infra}}{\omega} \sim \frac{c_1 + c_2 N_{AP}}{\omega} = c_2' + \frac{c_1'}{\omega} \qquad (12.4)$$

For a wireless system the fixed cost (of providing rudimentary coverage) c_1 is small and the capacity (i.e., the number of customers the system is capable of serving) is virtually proportional to the cost of providing it. The second term in (12.3) is therefore negligible and the cost per served subscriber is almost constant. This is in contrast to fixed access infrastructures that have a high initial cost to provide any service at all whereas the marginal cost of connecting another user is usually small. Another way of understanding this is to note that in a wireless system there is virtually no equipment or parts that are for the exclusive use of a single user. All of the infrastructure is shared. This is all illustrated in Figure 12.3. In the graph it can be seen that for low user traffic densities, as one would find in rural and suburban environments, cellular technology provides the cheapest solution. In city centers where the infrastructure can be shared by many, the fixed access costs are more favorable. Of course, also in these environments, the consumer would be interested in paying a premium for the mobility.

The cost structure of cellular telephony systems has been ideally suited for its introduction in the early 1980s. New operators with rather limited capital base were allowed to enter the market with rather moderate infrastructure investments. A minimal set of base stations barely covering the most densely populated areas and the major highways was the entry ticket. Due to the large demand, the capacity of the typical cellular system was soon exhausted. On the other hand revenues were substantial allowing the operators to invest in more base station, increasing the revenues, increasing investments in a fast moving upward spiral. The initial drive was of course aimed at providing mobile services, but at the end of the 1990s, new wireless mobile subscriptions actually surpassed the number of fixed subscribers in several of the pioneering countries. Although tariffs are still substantially higher for wireless mobile services, they have become extremely popular.

Although the regulators are easing their grip on the telecommunications market, there are still significant barriers for incumbent new actors in the wireless field. One is the strict regulation of the frequency spectrum, the other is due to standardization. In fact these two phenomena are tightly coupled. Although the national states are bound by treaties within the International Telecommunication Union (ITU), prescribing the use of the spectrum, they maintain sovereignty over which national actors should be awarded rights to use the spectrum. Licensing is the method by which national frequency authorities give the licensees exclusive rights to use a certain part

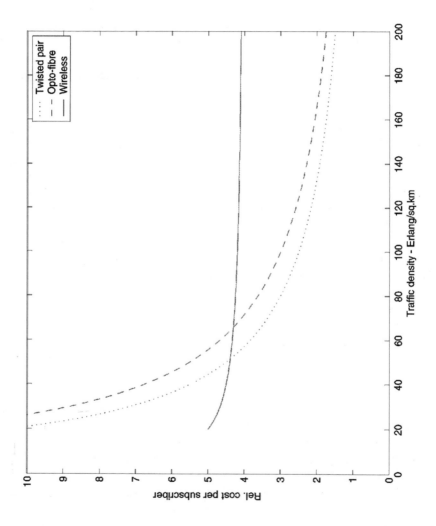

Figure 12.3 Cost per subscriber for various access techniques for access to the public telephone network.

of the spectrum for a certain purpose. The regulator seeks (from a public interest perspective) to organize the use of the spectrum as efficiently as possible, and to avoid unwanted interference between different services and users. In most licensing agreements the licensee is also required to use equipment and systems that are in compliance with certain standards. In the early days of the ITU (1950–1970), the organization also dominated the standardization process. Since the late 1980s the lead roles are now played by the equipment and systems manufacturers. Their organizations, like ETSI (Europe), ARIB (Japan), and TIA (USA) have become the leading standardization bodies.

Standards, in their original design, were necessary tools to create a mass market of terminals to bring prices down to an affordable level for the consumer. From the 1990s onwards, standardization also plays an important role in the business strategies of the mayor manufacturers [9]. By strongly influencing the standards, the big players can create system designs with high complexity, which effectively lock out smaller manufacturers and alternative technology, thus creating an oligopoly situation. The standardization process has become yet another barrier for incumbent players, and is in this respect not necessarily beneficial to the consumer. Nowadays, considerable interest is put into the development of systems that operate in so-called unlicensed frequency bands. In these bands the user is allowed to operate all kinds of equipment and systems fulfilling minimal technical requirements. The regulator provides no guarantees regarding interference from other (legal) users of the band. Examples of such systems are WLAN systems, for example, as specified by the IEEE 802.11 or the Hiperlan-II standards.

12.3 Cost Models for Wideband Wireless Infrastructures

As discussed in the previous section, today's mobile communication systems that are primarily designed to provide cost efficient wide area coverage for users with moderate bandwidth demands, have evolved very successfully. In the past few years, wireless data services have attracted increasing attention. Boosted by the massive use of popular Internet-based services such as email and the World Wide Web, wireless wideband networks capable of supporting mobile multimedia services are being planned. These services require (at least bursts of) high bandwidths. It is not expected that future users will be willing to sacrifice functionality for the added value of mobility—since expectations are that most telecommunication devices will be used by mobile users (and hence they will be wireless). An important requirement has been postulated for these wideband services—they cannot be substantially more expensive than the voice services offered today. If there is to be a widespread use of wideband multimedia services, it is widely accepted that the tariff structure cannot be proportional to the bandwidth provided in the way they are today. In wireline systems, we have, due to the introduction of optical fibers, experienced a sharp drop in communication costs measured per transmitted bit. The result is that additional bandwidth can be provided at almost zero marginal cost as soon as the fiber has been installed. However, wireless schemes have a much lower fixed installation cost. It is therefore not obvious that the success story of mobile telephony can be extrapolated, since the capacity measures, the number of users, and the bandwidth provided, are not directly interchangeable in a wireless environment. The key question remains: can wideband wireless multimedia services be provided at reasonable prices, that is, close to the cost of a telephone call today? Other limiting factors include mobile equipment power consumption, which increases linearly with the bandwidth provided. This section we will probe a little deeper into this issue. Some simple models will be used to analyze the cost structure of the two types of access architectures—a universal coverage scenario offering wideband services at all locations and a hot spot scenario where full-rate service is offered only in limited geographical areas. Both infrastructural costs and spectrum licensing costs are included in the analysis.

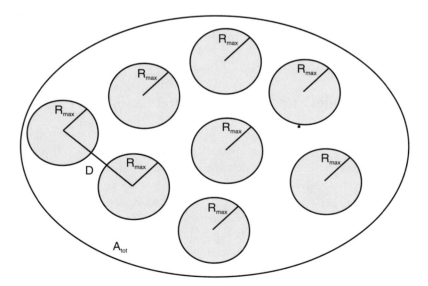

Figure 12.4 Service area and notation.

Consider a wireless network covering a certain service area, A_{tot}. In this area N_{AP} wireless access ports (base stations) are dispersed. Customers are randomly dispersed according to a 2-D Poisson process with parameter ω (users/area unit). A total of W_{sys} total bandwidth is available. This furthers the simplifying assumption that every user requires the same bandwidth W_{user}. The mobiles use a transmitter power of at most P_{batt}, giving them a maximal range of R_{max}. This corresponds to the maximal distance where the signal-to-noise requirement can be met in the absence of interference from other users in the system. Assuming that the path loss is proportional to the α-th power of the distance yields

$$P_{rx} \geq \frac{cP_{batt}}{R_{max}^a}$$

where α is typically the range 2–4 ($\alpha = 2$ corresponds to line-of-sight communication paths). Since the received power required to maintain a certain signal-to-noise ratio is proportional to the user bandwidth:

$$R_{max} \leq c_m \left(\frac{P_{batt}}{W_{user}}\right)^{1/\alpha} \tag{12.5}$$

where c_m is a constant dependent on the actual transmission system (modulation, coding, and so forth). Now, assume that the system is designed to let each access port provide service to mobile terminals in each immediate surrounding area. The base station service areas (cell) is approximated by a circles of the same radius R, which yields a cell area

$$A_{cell} \approx \pi R_{max}^2$$

For uniformly dispersed access ports, *seamless (or full) coverage* is achieved when

$$N_{AP} A_{cell} \approx \pi R_{max}^2 > A_{tot}$$
$$N_{AP} > \frac{A_{tot}}{\pi R_{max}^2} N_{AP}^* \tag{12.6}$$

that is, the maximum range R_{max} exceeds the critical cell radius R_0 given as

$$R_{max} > \sqrt{\frac{A_{tot}}{\pi B}} = R_0(N_{AP}) \qquad (12.7)$$

Assume that the transmission system is capable of handling interference from other users such that the available bandwidth may be reused in every K_0-th cell (corresponding to a reuse distance of D_0 as indicated by Figure 12.4). In this case the design is such that the entire service area is covered by cells. The analysis will now attempt to determine these costs for a given traffic load (number of customers) as function of the various system parameters.

Using the standard approach from Chapter 4, the maximal number of users accommodated by the system can now be computed as

$$N = N_{AP} \min\left(\frac{W_{sys}}{K_{eff}W_{user}}, \omega \pi R_{max}^2\right) \qquad (12.8)$$

where the first argument of the min-function corresponds to the case when interference limits the capacity. In the second case, the range of the terminals is so small that the user demand, expressed as ω users/unit area, limits the total population served by the system. The *critical demand* ω^* below which the system will be demand limited, can be computed by combining (12.6) and (12.8)

$$\omega^* = \frac{W_{sys}}{\pi(c_m)^2(P_{batt})^{2/\alpha}K_{eff}(W_{user})W_{user}^{1-2/\alpha}} \qquad (12.9)$$

K_{eff} denotes the *efficient reuse factor*. For seamless coverage of the entire service area, the reuse factor was assumed to be K_0. However, for systems with partial coverage, as illustrated in Figure 12.4, the spectrum may be reused more efficiently due to the unused buffer areas surrounding the base stations. It can easily be shown that the efficient reuse factor can be computed as

$$K_{eff} = \begin{cases} 1 & R_{max} < R_0/\sqrt{K_0} \\ K_0(R_{max}/R_0)^2 & R_0 = R_{max} = R_0/\sqrt{K_0} \\ K_0 & R_{max} > R_0 \end{cases} \qquad (12.10)$$

K_{eff} is a non-decreasing function in R_{max} and thus a non-increasing function of W_{user}. For the (from an operator point of view favorable) high demand, full coverage case, that is, ω large and $N_{AP} > N_{AP}^*$, (12.5) can be inserted into (12.8) to solve for the number of access ports N_{AP} that will be required to achieve a certain number of users served:

$$N_{AP} = N\frac{K_0 W_{user}}{W_{sys}}$$

Inserting this into the cost model I (12.2) yields:

$$C_{infra} = c_1 + c_2 N_{AP} K_0 \frac{W_{user}}{W_{sys}} \qquad (12.11)$$

As can be seen in the expression above, the infrastructural cost grows linearly with the number of terminals supported for any given fixed user bandwidth. This fact, as was noted in the previous section, is one of the prime success factors of the wireless community. The revenues of operator, which are proportional to the number of customers) grow (roughly) at the same rate as the investments are made. On the other hand, and what is more disturbing, the costs grow linearly with the user bandwidth as well. For the other cases, the situation is somewhat more complex and a numerical example is used to illustrate the behavior.

In the example, the transmitter (battery) power is set to some (interesting) low value such that almost full coverage is achieved from a single access port at very low bandwidths, whereas only very poor coverage is achieved at the highest data rates considered. Since the cost functions contain a number of constants no absolute numerical values but rather relative costs and bandwidths are presented. The reader should be able to reconstruct these graphs up to some constant. In all examples in this section the parameters $K_0 = 9$, $\alpha = 3.5$ have been used.

Figure 12.5 shows the relative cost per user (C_{infra}/N) for cost model I as function of the user bandwidth for some different demands ω. Studying the lower graph corresponding to $\omega = 100$, shows the linear growth in cost expected from the previous section for the low bandwidths where seamless coverage can be achieved. For W_{user} above 30, full coverage is no longer possible and the cost curve drops away from the linear behavior. If ω is decreased, the number of users served becomes demand limited, and the cost per user increases since there are less users to share the cost of the infrastructure.

Figure 12.6 shows the dependence of the cost on the available system bandwidth for cost model I. As can be expected from (12.11), the cost per

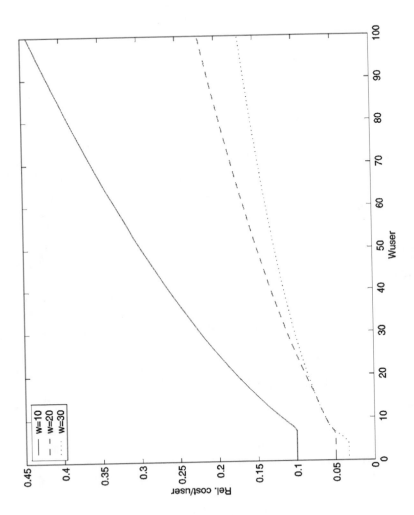

Figure 12.5 Relative infrastructure cost as function of user bandwidth provided for various user demands ω. ($K_0 = 9$, $W_{sys} = 1$, $\alpha = 3.5$, $N_{AP} = 100$).

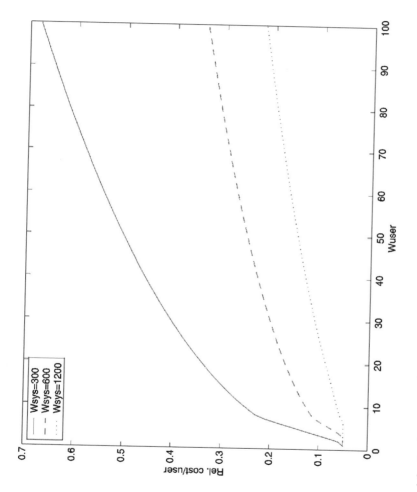

Figure 12.6 Relative infrastructure cost as function of user bandwidth provided for large user demands ($\omega = 100$) for varying available system bandwidth ($B = 100$).

user drops when the system bandwidth increases. It may not come as a surprise that if the spectrum was free of cost, an operator should try to acquire as much spectrum as possible to lower his costs. Applying cost model II, however, would penalize high system bandwidths by adding a constant term proportional to the system bandwidth to each curve. This would for instance shift the curve marked $W_{sys} = 4$ up more than the curve marked $W_{sys} = 2$ creating an intersection. Finding the optimum system bandwidth that an operator should bargain for is now no longer trivial.

How many access ports should an operator deploy in a wideband system? Figure 12.7 sheds some light on this matter. Starting with a very large number of access ports ($N_{AP} = 250$), we see that we achieve full coverage and the cost function is again linear in W_{sys}. For lower access port densities ($N_{AP} = 100$ and $N_{AP} = 25$), we see the characteristic knee where the link budget no longer allows full coverage and the curves break away from the linear behavior. For very low access port densities ($N_{AP} = 5$), this is not even noticeable in the graph and we find ourselves in the partial coverage situation for almost all user bandwidths. The set of curves suggest there is a optimum number of access ports for each user bandwidth.

Figure 12.8 illustrates the other side of the story, the total expected operator revenues. Here, the number of users served per access port is shown as function of the bandwidth for the same number of access ports as in the previous graph. Again one may note the rapid drop in the number of served users when we have full coverage ($N_{AP} = 200$) as compared to partial coverage systems.

The results again corroborate the initial assumption that in a wireless system with full coverage and a sufficient user demand, the infrastructure cost will rise linearly with the bandwidth. However, a system providing only partial coverage shows a much more favorable cost structure. The results show that for each offered user bandwidth and user demand there exists a number of base stations which minimize the cost of service provision per user, which (mostly) yields systems with only partial coverage. In order to achieve reasonable transmission costs for wireless multimedia services, hot-spots covering wideband schemes seems to be the only way to go.

An important issue is the question whether the users would accept that only partial coverage is provided by the operator or if substantial losses will would be incurred in this case. Hierarchical architectures, combining low to moderate rate universal coverage with wide-band hot-spot coverage could mitigate some of these problems. The implication of the results is to use multimedia terminals and services tolerant to variable data rates and commu-nication quality not only as proposed in a transient introduction phase, but

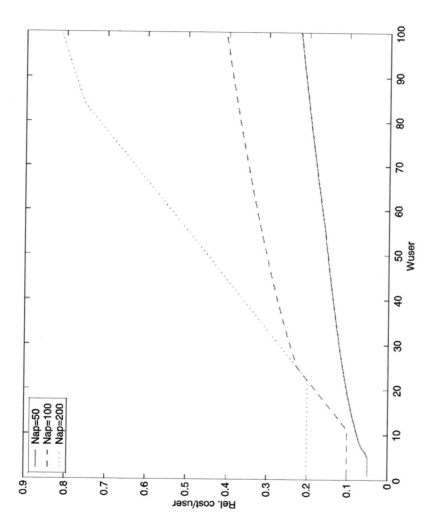

Figure 12.7 Relative infrastructure cost as function of user bandwidth provided for various numbers of base stations ($W_{sys} = 1$, $\lambda = 50$).

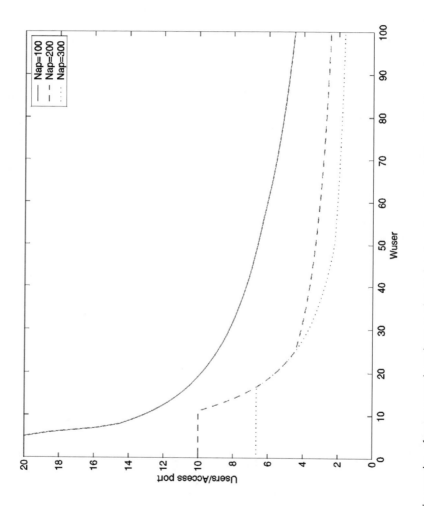

Figure 12.8 Relative number of users served per base station as function of user bandwidth provided for various numbers of base stations ($W_{sys} = 1$, $\omega = 50$).

also in the long-term perspective. Although the simplifying assumption that all users use the same bandwidth was made, similar results would also be obtained for varying user bandwidths if W_{user} is interpreted as the average data rate.

Problems

12.1 In a cellular system the path gain from an access port to a terminal follows the expression (see Appendix A)

$$G(r) = C - 35 \lg(r)$$

The system is designed to provide digital cellular telephony at an efficient data rate of 10 Kbps. At the design range, 5 km, where the system achieves full coverage, the signal-to-noise ratio, E_b/N_0, is 15 dB.
   a) Determine the area coverage (in percent) at 50, 200 Kbps, and 1 Mbps data rate.
   b) How many more access ports are required to achieve full coverage for the data rates in a)?
   c) How many more access ports are required to achieve 50% coverage for the data rates in a)?

12.2 In the system in Example 12.1, 5 MHz total system bandwidth is available. For a 50 Kbps user data rate and full coverage, how much lower would the cost be (using model I) if an additional 5 MHz spectrum was acquired? Repeat the calculation for the 200 Kbps case?

12.3 In the system in Example 12.1, calculate and plot the average power consumption of the terminals as function of the infrastructure cost. Use terminal power consumption and the number access points of the telephony system as reference. Plot the figures for 50, 200 Kbps, and 1 Mbps.

References

[1] Wheatley, J. J., "World Telecommunication Econonomics," IEEE, 1999.
[2] Arnbak, J. C., "The European Revolution of Wireless Digital Networks," *IEEE Comm. Mag.*, Sept. 1993.

326 Radio Resource Management for Wireless Networks

[3] Arnbak, J. C., "Personal Wireless Systems: Trends in Mobile Network Technology and Regulation," *IEEE International Conference on Personal Wireless Communications*, 1996.

[4] Arnbak, J. C., "Economic and Policy Issues in the Regulation IF Conditions for Subscriber Access and Market Entry to Telecommunications," *in* Korthals Altes et al., *Information Law Towards the 21st Century*, Boston: Kluwer, 1992.

[5] Haberkorn, R. A., and P. E. Nikolich, "Driving Forces in Wireless Datacommunications," *New Telecom Quarterly*, 1Q96, 1996.

[6] Morisson, E., "The Economics of Wireline vs. Wireless Telephony," *New Telecom Quarterly*, 3Q95, 1995.

[7] Shaw, J. K., "Wireless Communications and Technology Substitution: What S-Curves Reveal About Pending Cellular Competition," *New Telecom Quarterly*, 2Q96, 1996.

[8] Zander, J., "On the Cost Structure of Future Wideband Wireless Access," *IEEE Veh. Tech. Conf., VTC '97*, Phoenix, AZ, May 1997.

[9] Gessler, F., "The Development of the DECT Standard—An Example of Standardization in Wireless Communications," *Lic. Thesis*, KTH IEO R 2000-05, May 2000.

Appendix **A**

Propagation Models for Wireless Networks

The Maxwell equations together with the wave equation are a set of partial differential equations that, at least in principle, provide an exhaustive solution to all problems in electromagnetic fields and thus in radio propagation. Unfortunately, due to the massive complexity of these solutions, they are only of practical use in certain, simple and highly idealized cases. In most practical cases, the richness of details usually prohibits the accurate prediction of different propagation phenomena. The Maxwell equations thus provide little intuitive understanding of real-life problems. Instead, radio engineers tend to use simpler models that only give a coarse description of the physical phenomena, when viewed from a system perspective. In the following, a few of these simplified models that will be used in different parts of the book are introduced.

Distance Power Loss

Figure A.1 illustrates the received power at some terminal moving away from the access point in some given direction. The terminal is currently located at distance r. Classically, one has identified distance dependence consisting of the product of three components, corresponding to three different granularities in the distance r: the distance dependent average path gain, the shadowing gain and the multipath gain. The received power may thus be written as

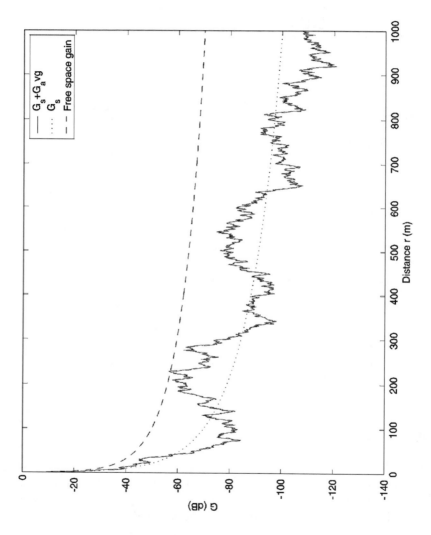

Figure A.1 Distance dependence of received power (link gain); $\alpha = 3$, $\sigma_s = 10$ dB, $a = 0.99$.

$$P_{rx} = P_{tx} G_{avg} G_s G_{mp} \tag{A.1}$$

or, in dB scale:

$$\overline{P}_{rx} = \overline{P}_{tx} + \overline{G}_{avg} + \overline{G}_s + \overline{G}_{mp} \tag{A.2}$$

Starting with the average path loss, the distance dependence of the global average G_{avg} can be modeled with a simple exponential expression

$$G_{avg} = \frac{C}{r^{\alpha}}$$

or

$$\overline{G}_{avg} = \overline{C} - 10\alpha \lg r$$

In its second (dB) form this model is usually referred to as the Okumura-Hata model [1]. The exponent α is empirically determined from measurements for different terrain types. A simple extreme case is free-space propagation ($\alpha = 2$) [2]. In terminal communications α-values between 3–5 have been used, where the low end corresponds to rural or suburban terrains. The higher values correspond to urban environments with high-rise buildings. We may note that also the plane-earth model is covered by this model, where $\alpha = 4$ corresponds to the case where the transmitter and receiver are far away from each other compared to the antenna elevations. The constant C is mainly dependent on antenna parameters such as antenna gain, elevation, etc.

Shadowing

Except for smaller objects obstructing the LOS-path between transmitter and receiver, there may also be major terrain obstacles, such as hills, large buildings, and so forth. A terminal moving behind such an object will be subject to shadowing. Due to the diffraction, waves will to some extent enter the geometric shadow region. At frequencies above 300 MHz, the amount of diffracted energy is quite low and shadows will thus be rather distinct. As the terminal moves through an environment where it will be (partially) shadowed from direct waves and collections of reflected waves, the received signal level will fluctuate. The phenomenon is called shadow fading. The signal level variations will depend on the relative position of the terminal to the shadowing objects. Since these objects may have considerable physical dimensions (hundreds of meters), it may take some time for the terminal to move out of a shadow region. The shadowing gain can be rather accurately estimated, given the exact geometry of the terrain, buildings, and obstacles in the surroundings of the transmitter and the receiver, by various diffraction models. Such techniques are, for instance, in various planning tools for cellular systems based on Geographical Information Systems (GIS), where height contour maps and terrain characterizations form the basis for rather accurate propagation prediction. Since the finer details of the terrain are never exactly known in practice (the maps have typically resolutions in the order 50–100m), a residual error will remain in all these predictions. The simplest prediction model is a purely stochastic model where the distance based average gain above is used. A common model for these signal level variations (or prediction error) is the log-normal distribution characterized by the probability density function

$$p(\overline{G}_s) = \frac{1}{\sigma_s \sqrt{2\pi}} e^{-\frac{(\overline{G}_s - \overline{G}_{avg})^2}{2\sigma_s^2}} \qquad (A.3)$$

In this case the log-standard deviation may be in the order of 8–12 dB, dependent on the terrain roughness. When more sophisticated prediction schemes, using the actual terrain geometry, are used, this value may be considerably less.

Combining the distance loss and the shadow fading, we have in decibel-scale:

$$\overline{G}(r) = 10 \lg(G(r)) = X(r) + C - 10\alpha \lg(r) \qquad (A.4)$$

where $X(r)$ is a zero-mean Gaussian random process. The log-normal shadow fading will certainly be correlated such that $G(r)$ takes similar values for nearby distance values r. A simple model for this correlation is the first order model proposed in [3]. In the model it is assumed that the autocorrelation between dB-samples of the signal level taken at sample rate $1/T$ in a mobile terminal moving at speed v is given by

$$R_X(k) = \sigma^2 a^{-|k|}$$

$$a = \epsilon_D^{vT/D}$$

where ϵ_D is the correlation of two shadow fading gains $X(r)$ and $X(r + D)$.

Multipath Fading

When looking at the fading process in a more microscopic scale than given by Figure A.1, the fast multipath fading is seen. Whereas the shadow fading is quite slow (10–100m), the multipath fading, which is dependent on phase differences between wave components, is very rapid (wavelengths, i.e., 0.5–1m). This is a typical narrowband signal phenomenon, when the delay of the multipath components is less than the symbol duration of the transmitted signals. For wideband signals, the received power fluctuations have considerably lower amplitude (the signal waveform is distorted due to time-dispersion instead). The reader is referred to [2, 4] for a more detailed analysis. Here, only the main results are stated. For narrowband waveforms where the received signal is the sum of many reflected signal components, the central limit theorem is used to derive the statistical properties of the received signal complex envelope [1, 5]. Assuming that the received signal consists of N components, it can be written as

$$z(t) = \{\xi(t)e^{-j\theta(t)}\} = \sum_{i=1}^{N} x_i(t) + jy_i(t) = \sum_{i=1}^{N} x_i(t) + j\sum_{i=1}^{N} y_i(t) = x(t) + jy(t)$$

where $x(t)$ and $y(t)$ become uncorrelated Gaussian processes, provided N is large and that the statistical properties of the different incoming components are the same. Computing amplitude and phase of $z(t)$ yields

$$\xi(t) = \sqrt{x^2(t) + y^2(t)} \tag{A.5}$$

$$\theta(t) = \arctan\left(\frac{x(t)}{y(t)}\right) \tag{A.6}$$

By substituting (A.6) into the joint probability density function (2D-Gaussian), we get the pdf:s of $\xi(t)$ and $\theta(t)$ as

$$p(\xi, \theta) = \frac{\xi}{2\pi\sigma^2}e^{-\xi^2/2\sigma^2} \tag{A.7}$$

$$p(\xi) = \frac{\xi}{\sigma^2}e^{-\xi^2/2\sigma^2} \tag{A.8}$$

The (marginal) distribution of ξ is usually denoted the Rayleigh-distribution. This is why the multipath fading in some cases is somewhat loosely referred to as "Rayleigh fading." The process $\theta(t)$ is, according to (3.21), uncorrelated with ξ and has a uniform distribution over the interval $(0, 2\pi)$. The two first moments of ξ may now be evaluated:

$$E[\xi] = \sigma\sqrt{\frac{\pi}{2}} \tag{A.9}$$

$$\text{Var}[\xi] = 2\sigma^2\left(1 - \frac{\pi}{4}\right) \tag{A.10}$$

Further, we may evaluate the statistical properties of the instantaneous power of $\xi(t)$, $\gamma(t) = \xi^2(t)$. This latter stochastic variable is exponentially distributed, that is,

$$p(\gamma) = \frac{1}{\gamma_0} e^{-\gamma/\gamma_0} \tag{A.11}$$

whereby we

$$\gamma_0 = E[\gamma] = E[\xi^2] = 2\sigma^2 \tag{A.12}$$

denote the average (expected) power of the signal. The assumption that has been used here is that all reflected components have (statistically) the same energy. This is typically the case in non-line-of-sight situations where the received signal is the sum of a large multitude of scattered waves, all caused by the same type of physical phenomenon in roughly the same geometric conditions, and in a mobile radio environment, when a fixed base station is communicating with a terminal roaming in a heavily built-up area. A large number of reflected waves incident from all angles would contribute to the received signal in this situation. If, in the latter case, there is also a strong component due to some other phenomenon than reflections in the surrounding buildings, that is, a steady direct line-of-sight component, we have to modify our expression as

$$z(t) = \eta + x(t) + jy(t)$$

Again using (A5, A6) and exploiting the fact that x and y are independent Gaussian variables, the joint probability density functions of ξ and θ can be computed. One of the results, the marginal distribution of ξ, is

$$p(\xi) = I_0\left(\frac{\xi\eta}{\sigma^2}\right)\frac{\xi}{\sigma^2}\,e^{-(\xi^2+\eta^2)/2\sigma^2} \qquad\qquad (A.11)$$

I_0 is the zero-order modified Bessel function and the parameter η denotes the amplitude of the dominant, constant signal. This distribution is usually called the Nakagami-Rice-distribution or just the Rice-distribution.

References

[1]   Hata, M., "Empirical Formula for Propagation Loss in Land Mobile Radio Services," *IEEE Trans. Veh. Tech.*, Vol. VT-29, No. 3, Aug. 1980.

[2]   Ahlin, L., and J. Zander, *Principles of Wireless Communication, Studentlitteratur*, 1997.

[3]   Gudmundson, M., "A Correlation Model for Shadow Fading in Mobile Radio," *Electronic Letters*, Vol. 27, Nov. 1991, pp. 2146–2147.

[4]   Proakis, J. G., *Digital Communications*, New York: McGraw-Hill, 1995.

[5]   Jakes, W. C., et al., *Microwave Mobile Communications*, New York: Wiley & Sons, 1974.

Appendix **B**

Simulation Models

Modeling Time

There are two different ways that time can be modeled in a simulator for cellular systems.

In the event driven model, everything that happens in the model is associated with a point in time. This can, for example, be the arrival of a new mobile, the termination of a call, or updating the position of a mobile. All these events are placed in an event queue, and they are performed one event at a time.

In the discrete time step model, the cellular system is only studied at specific (regularly spaced) time instants. Between time instants many mobiles may have moved, new calls may have been created and others may have been terminated. In general the whole system changes between each time instant.

There are advantages and disadvantages with both methods. The advantage of the event driven model is that it is easy to accurately model the time events that take place. There is no sampling error. Also if there are long periods when no events take place, those periods do not require any computations.

The advantage of discrete time steps is that the whole system can be handled at the same time. It is practical to represent the state of a system in vectors and matrices. These can then be treated quite efficiently using mathematical software (e.g., MATLAB). This type of representation is also suitable for computation on parallel computers. This means that the simulations can run faster than for an event driven system where each event has to be treated separately.

The difficulty in implementing the simulation program also plays a role when deciding which paradigm to use. In general the implementation of a simulator with discrete time steps is easier than implementing an event driven simulator.

The paradigm chosen for time representation has a large impact on the design and implementation of a simulator. The RUNE simulator is based on discrete time steps to increase simulation speed and also to simplify implementation and debugging of the simulation programs.

Cell Layout—Service Area

When creating a model of a cellular system, the aim is to produce results that closely resemble the results that would be achieved in a system in the real world. Real systems can contain thousands of cells. It is of course possible to create a model that contains all of the cells. The major drawback is that simulating all of the cells is extremely time consuming.

To reduce the computational load of the model, a small set of cells are chosen. These cells should be representative of cells in the real system. Thus it is assumed that the results obtained by simulation can be applied in a practical cellular system. In a practical system other cells surround most of the cells. That means that in the model, a set of cells that is surrounded by other cells should be used to get representative results.

One way to make sure that the cells are surrounded by other cells is to first create the cells that are under study. Then these are surrounded with more cells from which no results are required. The problem with this approach is that even though no results from the surrounding cells are used, the same effort is put into simulating the surrounding cells as is put into simulating the internal cells. The net result is that the simulation effort is large.

Another way to make all the cells appear as if they were placed in the middle of a cellular system is to use a wraparound technique. The cells under study are placed in a rhombus. This is then stitched together so that the top meets the bottom and so that the left side meets the right side. This creates a torus-shaped surface that the cells are placed on. It is worth noting that this surface has no borders. This means that all cells will have neighbors on all sides. It also means that there will be an infinity of straight paths between any two points on the torus.

Another way to view the torus-shaped model of the world is to view it as a parallelogram that is repeated infinitely many times upwards, down-

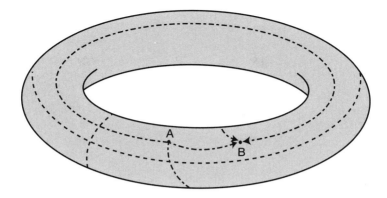

Figure B.1 The cellular system is placed in a rhombus, which is stitched together to a torus along the dashed edges. There is an infinity of paths between A and B. Three examples are shown.

wards, and to the sides. However, it is worth noting that the torus surface is finite. Thus the cellular system is not infinite, but there are no borders.

Propagation Modeling

The path-loss between a mobile and base station depends on many things in a practical system. But in order to be able to build a model that is simple enough to handle, only a few issues are considered. The propagation loss is modeled as a sum of three or four components: the antenna pattern, the distance dependent fading, log-normal shadow fading, and sometimes Rayleigh fading.

Base Station Antenna Diagram

The antenna at the base station may contribute to losses or gains depending on the antenna diagram and the direction to the mobile. The base station antenna is modeled as different gains in one-degree sectors.

Mobile stations are considered to have no directional antenna. Although it is possible to have a directional antenna at the mobile station, it is usually too large physically to fit nicely onto a handheld mobile.

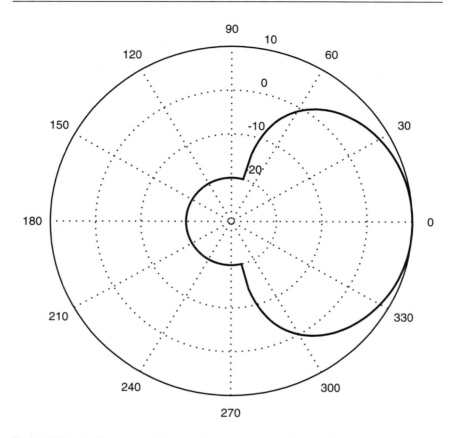

Figure B.2 Typical antenna diagram of an antenna that is used for three sector sites. The gain in the main direction is 7 dB.

Distance Dependence

The distance dependent fading can be modeled as:

$$G = \frac{C}{r^\alpha}$$

The constant C is used to model the effects of antenna size and other physical parameters. It can be interpreted as the path gain at a distance of one meter from the transmitter antenna. At this close distance, free space propagation is assumed. For a GSM system at 900 MHz with isotropic receiver and transmitter antennas, the path gain is approximated as

$$C = \frac{\lambda^2}{(4\pi r)^2} \Rightarrow C \approx -31 \text{ dB}$$

The α parameter determines how much the power decays as a function of distance from the base station. For free space propagation this parameter is 2, and in a typical urban environment this parameter ranges from 3 to 4.

It is easy to see that if the mobile station is very close to the base station, that is, when r is close to 0, G will extend to infinity. In this case there is a large discrepancy between the model and reality. In reality the near-field effects dominate close to the antenna. In addition it is not possible to come close to the base station antenna in a practical system since it is usually mounted in a tower of some kind. However, in the model, there is nothing to prevent a mobile from ending up at the same position as the base station. To avoid this it is assumed that all mobiles that are closer to a base station than a specified distance are located at that distance from the base station. This gives us the pathloss curve in Figure B.3.

Log-Normal Shadow Fading

There is a log-normal component of the propagation loss that models things like shadowing behind obstacles, losses due to passage through walls, and so forth. The log-normal shadow fading can be modeled as (see also Appendix A)

$G = 10^X$, where X is normal distributed with mean 0 and variance σ.

One way to model the log-normal shadow fading would be to assign it a random value each time the pathloss is calculated. The drawback with this method is that the shadow fading does not have any correlation in space.

In the real world the shadow fading is correlated in space. If a mobile for example is behind an obstacle it will probably still be behind that obstacle if it only moves a small distance.

One way to make sure that the shadow fading is correlated in space is to use the distance a mobile has moved for determining how much the shadow fading should change. The drawback with this method is that if a mobile returns to the same position, the amount of shadow fading will not be the same.

The way to ensure that the shadow fading is correlated in space and that the amount of shadow fading will always be the same in the same

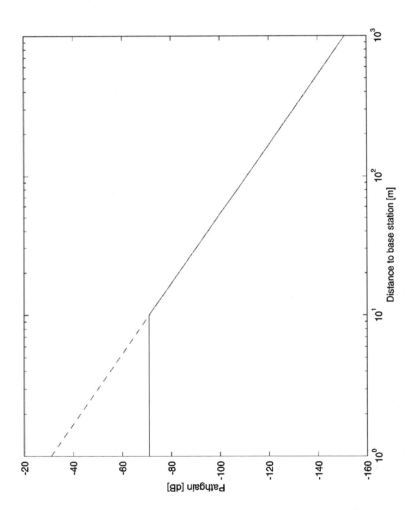

Figure B.3 The path gain between base station and mobile as a function of distance $\alpha = 4$. Mobiles closer than 10m are considered to be 10m distant from the base station.

position is to give each geographical point in the system a specific amount of shadow fading. The easiest way to ensure this is to create a data structure that contains the shadow fading for all points in the model. But this is impractical since it consumes a lot of memory.

To avoid the problem of memory consumption, a smaller map is created, and it is repeated many times over the system area. It is made so that the borders can be stitched together without abrupt changes in the gain values. By using a 2D-Fourier transform it can easily be ensured that the borders fit together. Finally, to find a value for the lognormal shadow fading between the points in the lognormal map interpolation is used.

In a practical system there is often a correlation between the lognormal shadow fading for the links between one mobile station and the base stations. For example when a mobile station moves down into a cellar, it will experience increased pathloss to all base stations. This is modeled as a correlation in the shadow fading between one mobile and the base stations. This is done by assigning one random fading component to each link G_{ij} and one fading component related to the mobile G_M. These are then added according to the expression:

$$G = \sqrt{\rho}\, G_M + \sqrt{1 - \rho}\, G_{ij}$$

From this it can be seen that if ρ is set to 1 the expression above will only contain the fading component related to the mobile. This means that the links between one mobile and all base stations will be subjected to the same lognormal shadow fading. This also means that both the carrier and the interfering signals fade by the same amount.

Rayleigh Fading

In some cases it may be desirable to model Rayleigh fading. This is essentially done in the same way as for lognormal shadow fading. A 2D map is created which is then tiled over the area. The borders here are also made such that the edges of the map fit together.

One problem however is that the Rayleigh fading varies quickly over distance. This means that the points in the map have to be quite dense to accurately model the fading. Typically one point is required every half wavelength. For a 900 MHz carrier frequency, this means that there has to be a data point every 15 cm. To achieve this the map either has to be very large or be repeated many times.

The problem with a large map is that it consumes a lot of memory for storage. The problem with repeating the map many times is that there is a possibility for cyclic behavior in the fading. If a mobile moves a distance that is equal to the map size it will experience the same fading at all times.

To avoid the problem with the cyclic behavior of fading, two maps for Rayleigh fading are used. These two maps have almost the same size, but not quite. Both are repeated over the system area. To obtain the fading in a point the sum of both maps are used. Since the sizes of the maps are a little bit different, there will be many repetitions of them before they are aligned again. This means that the distance between identical points becomes quite large and the problem with cyclic behavior can be avoided.

Traffic Generation

Generating traffic in a simulator based on discrete time steps can be a little bit tricky. Since the time in between time steps is not modeled, things may happen there that are missed. However if the time step is fairly small the probability that this will happen is usually negligible.

What happens in a real cellular system can be drawn as shown in Figure B.5. At time T_1 there are a number of active calls in the system. We then study what happens until time T_2. During that time some calls may arrive and be blocked. Other calls arrive and terminate again before the time T_2. Other calls arrive and they are still active at T_2. This is a very complicated model and simplifications are needed to make the model useful.

In order to determine which simplifications can be done, some design requirements are imposed on the model.

- Traffic build-up should be modeled fairly accurately when the time step is small.
- If the system is in a stationary state at a time instant it should also be in a stationary state at all times following that (unless traffic parameters change). For a system in a stationary state, the average number of users at each time instant should be the same and the average should be independent of the time step length.

Even though no traffic in reality is Poisson distributed, it is assumed that the traffic in the simulator is Poisson distributed. The reason is that it is mathematically convenient to handle. Another simplification made is that from a traffic generation perspective there is an infinity of channels. That

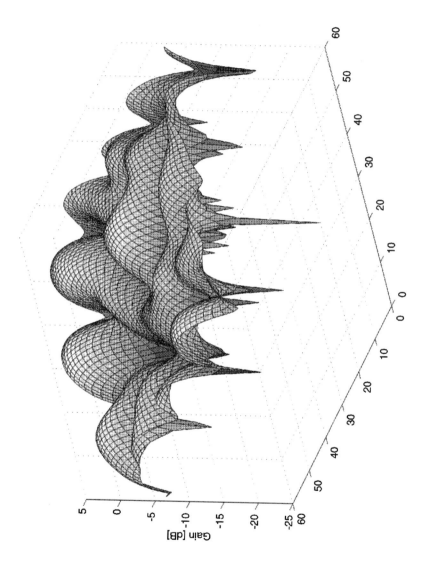

Figure B.4 A typical map describing Rayleigh fading.

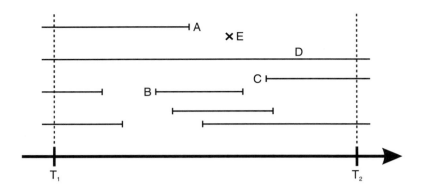

Figure B.5 In the time between two timesteps (T_1 and T_2) some calls are terminated (A). Some calls arrive and terminate (B). Others arrive and stay until the next timestep (C). A number of calls exist at T_1 and are not terminated during the period (D). Finally some calls arrive but are blocked (E).

way blocked users do not disturb the calculations. For small blocking rates this is a reasonable assumption, and it is usually those cases that are of interest.

In the model it is assumed that the average call duration is mean holding time (mht) long ($1/\mu$ = mht). In queuing theory, the arrival rate is commonly used (λ). However, it is impractical to use the arrival rate as an input parameter to the simulation. Instead it is preferred to use the traffic offered to the system (offtraff). If we disregard the blocking, this is the same as the average number of users in the system at the stationary state (N_0). From this, the arrival rate can easily be calculated as

$$\lambda = \text{offtraff/mht}$$

Of the users that are present in the system at time T_1, it is easy to determine which of them should leave. Since the service time is exponentially distributed, a simple random experiment will select the ones that should leave the system.

If there are on average N_0 users in the system at T_1 then at time T_2 on average $N_0 e^{-\mu(T_2-T_1)}$ users will remain. It can then be seen that if the average number of users is to remain constant adding on average $N_0(1 - e^{-\mu(T_2-T_1)})$ users is required.

The probability distribution of the number of new users that should be added to the system is complicated to determine. However assume that this is a Poisson distribution. For the cases when the time step is large or small this is a reasonable assumption.

To generate the number of new users the following procedure is used. First the average number of users to add is determined. Then a random experiment is performed to determine how many actually should be added.

A Poisson distributed random number is difficult to generate. However it can approximated by a binomial distribution. The average of a binomial distribution is np. If p is small this is a good approximation to a Poisson distribution. This is achieved by letting n be 100 times bigger than the average, which gives us a p of 0.01. This gives a good approximation to the wanted distribution.

Mobility

One easy way to model how mobiles move around is to let them perform what is called a random walk. Each mobile is moved a random distance in a random direction at defined times. Though this model is simple it is quite different from how mobiles move around in a practical system. In the real world a mobile generally keeps moving in the direction it has moved before. The typical example is mobiles moving along a straight road.

To overcome the problem with a random walk, each mobile is given a velocity as well as a position. In each time increment the mobile is moved according to the current velocity and the size of the time increment. The velocity of the mobile is also changed a little bit so that the mobile accelerates, slows down, and changes direction. Note that the velocity is a vector quantity, it has both a magnitude and a direction.

The velocity is updated according to the formula

$$\rho = e^{\frac{-dt \cdot a_{mean}}{v_{mean}}}$$

$$v_n = v_{n-1}\rho + \sqrt{1 - \rho^2}\, v_{mean}X$$

The ρ is the correlation of the velocity between time steps. If the time step or mean acceleration is large compared to the mean velocity that means that there will be little correlation between the velocity in each time step. The magnitude of X is Rayleigh distributed with mean 1 and a random direction. If there is no correlation between the time instants, it can be seen that the velocity will have a random direction each time. However if the correlation between time steps is large, the velocity will only change a little bit.

Since a rhombus represents the service area, but is really the surface of a torus, it is necessary to do something about the mobiles that move outside of the rhombus. The solution is to create wraparound for the mobiles as well. That means that if a mobile crosses the left side of the area, it is placed on the right side of the rhombus.

Appendix C

RUNE Tutorial

Computer simulation is one of the main tools in research and development of mobile radio communication networks. RUNE (Rudimentary Network Emulator) is a set of MATLAB functions that makes it possible to simulate a cellular network. RUNE consists of approximately 60 functions. These functions extend MATLAB to provide the necessary tools for simulating cellular systems.

RUNE provides functions that handle various aspects of a cellular system, that is, mobiles, base stations, propagation loss, interference, and mobility. RUNE also provides programs that contain complete simulators.

The RUNE functions can be used in many different ways. The programs can be used as they are to perform standard simulation tasks. It is also possible to write new scripts that use the RUNE functions in new ways. Finally the functions themselves can be modified to change the behavior of the simulator.

The aim of this tutorial is to provide an introduction to the most commonly used parts of RUNE. During this chapter we will create a small simulation program that simulates a cellular system, creates some mobiles in that system, and calculates the signal to interference ratio for these mobiles.

Creating Sites

The first thing a simulation program must do is to create cells. RUNE provides a function for this purpose. It is called *crecell*.

This function produces three data structures, xyb, fib, and rombvec. The first (xyb) contains the position of the base stations, the second (fib)

347

contains the direction of the antenna lobe, and the third (rombvec) contains the size of the rhombus that represents the service area of the system that is simulated.

In RUNE, positions are represented as complex numbers. The real part represents the east west direction and the imaginary part represents the north south direction. The reason for this is convenience. The functions needed for manipulating positions are already present in MATLAB, and they support complex numbers. One warning has to be raised though. The transpose matrix function in MATLAB (') does a complex transpose by default, which changes the imaginary part of the elements in the matrix. This means that the position is changed. This is not something that usually is intended. However there is a function for transposing the matrix without affecting the elements (.'). This is the function that should be used for transposing matrices that contain positions.

Try to create a few base stations by typing the following commands on the MATLAB command line:

```
[xyb, fib, rombvec] = crecell(1000,1,1,1,1)
plothex(xyb, fib)
```

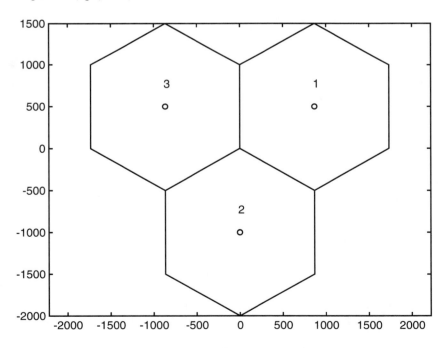

Figure C.1 The cellular system created by the commands above.

The first line will create three omnidirectional cells with a radius of 1,000m. Then the second line will create a plot of the three base stations. The positions for the base stations are put in the vector xyb.

The fib vector contains the direction of the main lobe of the base station antenna. For each base station position there is a corresponding direction of the main antenna lobe. The value is a vector that points from the base station to the middle of the cell. In this case the base stations have omnidirectional antennas. That means that the distance from the base station to the cell center is zero. This can be seen in the fib vector that contains only zeros.

The first parameter to crecell is the cell radius. It is easy to see from the figure that the cell radius is indeed 1,000m. The second parameter to crecell is the number of sectors per site. This can be one for omnidirectional antennas or three for sites that consists of three 120-degree antennas. Try to create a couple of base stations with directional antennas instead. Type the following commands:

```
[xyb, fib, rombvec] = crecell(1000,3,1,1,1)
plothex(xyb, fib)
```

This will create three sites with three antennas on each site. Note that the vector xyb contains nine positions. The result is three basestations that are located at the same position and each base station has an antenna pointing in different directions. This can be seen in the fib vector.

There are three more arguments to the function crecell. They are km, lm and ncluster. These parameters are used to set the reuse factor and the size of the system.

The creation of a cellular system is done in two steps. First one cluster is created. The cluster concept is closely coupled to reuse in cellular systems. One cluster contains a number of cells and all the cells use different channels. The next step is to copy the cluster $ncluster^2$ times and lay the clusters out next to each other.

Since the cluster is copied and the copies are laid out next to each other, it is important that the clusters fit together like a jigsaw puzzle. If three rhombuses are created that have the size km^2, lm^2 and $km \cdot lm$, and we put them together we get a piece that can be tiled together nicely. This means that each cluster will contain $km^2 + lm^2 + km \cdot lm$ cells.

To see how a cluster is built up, it is possible to use the function ex_crecell. Try the following:

```
ex_crecell
```

This will illustrate how a cluster with 36 cells is built. In Figure C.2 it is easy to see how the clusters fit together nicely. The parameters km, and lm can be input to the function for example when km = 2 and lm = 1:

```
ex_crecell(2,1)
```

Now start writing the simulation program. The simulation program should create a number of base stations. Then add a code that plots the cells. Create a function that contains the following code:

```
function tutorial

    % Create the base stations
    par.cellradius=1000;
    par.sps=1;
    par.km=1;
    par.lm=2;
    par.ncluster=2;

    [xyb,fib,rombvec]=crecell(par.cellradius,par.sps,par.km,par.lm,par.
ncluster);

    % Plot the system
    figure(1);
    plothex(xyb,fib);
```

Creating Mobiles

The next thing the simulation program should do is to create a number of mobiles.

In RUNE, mobiles are represented by a number of vectors in the first dimension (column vectors). Data belonging to one mobile is found at the same position in each vector. At this point, the vectors containing the number of the mobile (m), the mobile position (xym) and the velocity of the mobile (xyv). The names of the vectors can of course be anything, but out of convention they are called m, xym, and xyv.

The function responsible for positioning the mobiles and moving them around is called *mobmove*. The idea is that if a mobile does not have a position it is given one, and if a mobile already has a position it is moved a small distance. The reason for this is that mobmove often work with both mobiles that have been in the system some time and new ones simultaneously. When a mobile has not been given a position this is marked by setting the position in xym to nan.

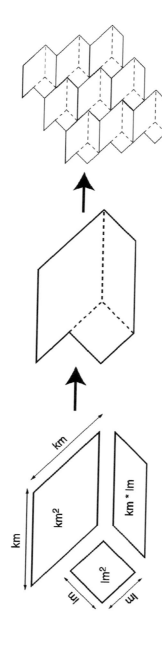

Figure C.2 Each cluster is made of three parts. The clusters are then duplicated and fitted together.

Create a few mobiles and give them a position by typing the following commands:

```
xym=nans(10,1)
xyv=nans(10,1)
m=(1:10).'
[xym,xyv]=mobmove(xym,xyv,10,5,10,rombvec)
plot(xym,'b*')
```

This gives the mobiles an initial position and an initial velocity. By using *mobmove* again the mobiles can be moved a little bit more. To see this, type the following commands:

```
[xym,xyv]=mobmove(xym,xyv,10,5,30,rombvec)
hold on
plot(xym,'r*')
[xym,xyv]=mobmove(xym,xyv,10,5,30,rombvec)
plot(xym,'g*)
hold off
```

Now the mobiles should have moved a little bit each time mobmove is called.

Mobmove takes a number of parameters. The first two are the mobile position and the mobile velocity. These vectors are updated each time mobmove is called. But mobmove takes four more arguments. The next two are the average speed in m/s and the average acceleration in m/s^2. These parameters are used when the mobile velocity is updated. The fifth parameter is the time step in seconds. The mobile moves a distance that is equal to the mobile velocity multiplied by the time step.

The last argument is rombvec. Rombvec contains two vectors that span the area containing the system. Mobmove needs to know how large the system area is for two purposes. The first is that mobmove should be able to spread the mobiles evenly over the service area. The second use is when a mobile already has been given position and is moving around in the service area. Even though the service area is represented as a rhombus, it is really a torus. Since the surface of a torus does not have any borders it is not possible for a mobile to move over the edge. But since we represent this area as a rhombus, something needs to be done when a mobile moves across the border of the rhombus. This is automatically handled by mobmove. When a mobile moves across one edge it is automatically moved to the opposing edge. For example if a mobile moves across the left edge it is moved to the right edge.

Now create the part of the simulation program that creates a number of mobiles and puts them into the system. Let the program create a specific number of mobiles (10). In many simulation programs the number of mobiles is a random quantity. It should be easy to extend the program to do this if necessary.

Add the following code before the plot part:

```
% Create new mobiles
par.vmean=10;
par.amean=5;
par.dt=10;

nmob=10;
xym=nans(nmob,1);
xyv=nans(nmob,1);
m=(1:nmob).';

[xym,xyv]=mobmove(xym,xyv,par.vmean,par.amean,par.dt,rombvec);
```

Add the following code after the plot part:

```
hold on;
plot(xym,'b*');
hold off;
```

Calculating Gain

The next thing required of the simulation program is to calculate the pathgain between all mobiles and all base stations. But before the function that does this *pathgain* can be called, some data structures for this function have to be created.

Each base station has an antenna that may be directional. A vector that contains the antenna gain in each direction in steps of one degree represents the antenna diagram of an antenna. The gain in this vector is given in dB. The vector is usually called lobevector.

RUNE is usually used with either omnidirectional antennas or with directional antennas that cover a 120-degree sector. An omnidirectional can easily be created. All that is needed is to create a vector that contains 360 zeros. Using `zeros(360,1)` can easily do this.

The directional antenna is a little bit tricky to create. Fortunately there is a function that does that for us: *sinclobe*. Try to create a directional antenna and plot the antenna diagram:

```
lobevector=sinclobe
lobevector=lobevector+40;
polar((1:360)*pi/180,lobevector')
```

The first line creates a vector with the antenna diagram in lobevector. Since the polar plot function that is to be used cannot handle negative values, we shift the values in lobevector so that they all become positive. Finally we create a polar plot of the antenna lobe. It is this type of plot that is normally used to show antenna diagrams. Note that the gain values have been shifted by 40 dB because of the second line.

As can be seen from the plot, the antenna has approximately −3.5 dB gain at 60 degrees from the main direction of the lobe. This is done so that the signal strength in the two closest corners and the corner of the cell farthest away from the base station is the same. Since the distance is two times that to the corner farthest away compared to the closest corner, we obtain the gain (given that α is 3.5).

$$G = \frac{(2R)^{\alpha}}{R^{\alpha}} = 10.5 \text{ dB}$$

To calculate the lognormal shadow-fading RUNE uses a map. The advantage with this is that the fading will always have the same value at specific position. The function that calculates the lognormal map is called *crelognmap*. It takes three arguments. The first two, xyb, and rombvec, are used to determine the size of the map. The third controls how rapidly the value of the shadow fading will change. Create a map and plot it:

```
[lognmap,mapvec]=crelognmap(xyb,rombvec,110);
mesh(lognmap)
```

The first output parameter contains the map itself. For each point it contains the fading value in dB. The mapvec contains two vectors of the size of the map. It can be viewed as the distance at which the map will be repeated. Note that the map will always contain approximately the same number of peaks and valleys. But it the correlation distance changes (the third input parameter) the size of mapvec will vary also. Check the values of mapvec and then create a new map with larger correlation distance.

```
mapvec
[lognmap,mapvec]=crelognmap(xyb,rombvec,1100);
mapvec
```

Note that the size of mapvec is increased by approximately an order of magnitude. An integer number of maps must fit within the system area to make the wraparound work nicely. Thus the value of mapvec is not a simple function of the correlation distance.

Now the gain between all base stations and all mobiles can be calculated. This is done by the function pathgain. It takes eleven input parameters. The best way to find out which is to browse the help text:

```
help pathgain
```

Calculate the gain for all links in the system we have created in the working environment:

```
g = pathgain(xym,xyb,fib,rombvec,-21,4,8,0.5...
lobevector, lognmap, mapvec)
```

The output matrix contains the gain values in dB. Each row corresponds to a mobile and each column corresponds to a base station.

Now it is possible to add the code to the simulation program that calculates the gain for all links. Add the following code to the simulation program:

```
% Create the antenna lobe
lobevector=zeros(360,1);

% Calculate the gain between bases and mobiles
par.attconst=-21;
par.alpha=4;
par.sigma=8;
par.raa=0.5;
par.corrdist=110;

[lognmap, mapvec] = crelognmap(xyb, rombvec, par.corrdist);
g=pathgain(xym,xyb,fib,rombvec,par.attconst,par.alpha, par.sigma,
par.raa, lobevector, lognmap, mapvec);
```

Allocating Channels

Now each mobile can be allocated a communication channel. There are many ways of determining which base station a mobile should communicate with and which channels should be used.

In the simulation program a simple strategy is used. A fixed channel reuse pattern is created to determine which channels each base station can use. Then the mobile is connected to the base station to which it has the highest pathgain.

There is a function *crechanplan* that makes a channel plan. Try to create a small channel plan:

```
obk=crechanplan(length(xyb),18,1)
```

This will create a channel plan with 18 channels for the nine cells created previously. The rows in obk correspond to a base station and the columns correspond to channels. If a matrix element is one it means that the channel can be used by the base station. For example base station 1 (row 1) can use channel 1 and channel 10.

The function *crechanplan* takes three input arguments. The first one is the number of base stations in the system. Usually this number is found by using length(xyb). The next argument is the total numbers of channels that the channel plan should have. In order to get a nice and symmetrical channel plan, each cell in a cluster has to get the same number of channels. For example if the reuse factor is $K = 7$ the total number of channels has to be a multiple of 7.

The last argument to crechanplan is the ncluster parameter. It simply tells crechanplan how many clusters the base stations are divided into. From this it can determine the reuse factor.

When the mobile is connected to a base station, new datastructures that represent the mobile are required. Two vectors are used to store the necessary information. The vector b contains the number of the base station that the mobile is connected to. The other vector, k, contains the number of the channel that the mobile uses to communicate. In the same way as for xym and xyv, nan is used to signify that the mobile is not connected to a base station.

There are three functions for connecting, handing over, and terminating connections to mobiles. These are assign, handoff and terminate. Here only assign and terminate will be used.

Whenever mobiles are assigned a new channel, or calls are handed off or terminated the obk structure is updated. If a mobile is assigned a channel and a specific base station the one in the obk matrix is set to zero to indicate that calls cannot be assigned to that channel in that specific base station any more. When the call is terminated the value in the obk matrix is set back to one again.

The input parameters to assign are fairly straightforward. The k, b, and obk matrices are input arguments and these are also returned in their updated version. In order to know which base station to connect to, the g matrix is also supplied. The fifth input argument, hand over margin, might require some more explanation though.

Sometimes it is not necessarily the case that the system should unconditionally connect to the strongest base station. Imagine that it is interesting to study handover in a cellular system. In order to study this we will need a couple of mobiles that are not connected to the strongest base station. These are the mobiles that should be handed over. To achieve this it is possible to create a system and move the mobiles around a little bit. The drawback with this method is that it may take a lot of computations to move the mobiles around. Another way to quickly generate some mobiles that should handover is to add a random quantity to the pathgains between a mobile and the base stations before we decide which link has the highest pathgain. This way there is a possibility that the mobile does not connect to the strongest base station. This is exactly what we wanted.

Assign provides this possibility. The fifth input parameter can be used to set the standard deviation of the "noise" that we add to the pathgain before it is decided which basestation is the strongest.

Make the datastructures k and b for the mobiles and assign the mobiles to channels.

```
k=nans(10,1)
b=nans(10,1)
[b,k,obk] = assign(b,k,g,obk,0)
```

Note that some of the ones in the obk matrix disappear.

Terminate works in a similar way to assign. The b, k, and obk data structures are sent as input parameters and are updated. The fourth input argument is a vector with the same size as k. If it contains a one, the call is terminated. Try to terminate the first call.

```
ter=zeros(size(k))
ter(1)= 1
[b,k,obk]=terminate(b,k,obk,logical(ter))
```

Note that the channel is returned to the obk matrix.

Now it is time to add some more lines to the simulation program that assign channels to the mobiles that have been created. Insert the following code in the simulation program.

```
% Create a channel plan
    par.kpc=5; % The number of channels per cell
    reuseFactor=par.km^2+par.lm^2+par.km*par.lm;
    obk=crechanplan(length(xyb),par.kpc*reuseFactor,par.ncluster);
```

```
% Create the necessary data structures and assign the mobiles to
% their channels
b=nans(nmob,1);
k=nans(nmob,1);
[b,k,obk]=assign(b,k,g,obk,0);
```

Calculating C/I

The final thing required of the simulation program is to calculate the carrier strength, the interference, and the C/I ratio both in the uplink and the downlink.

Before the calculations can be done it is necessary to know what the noise floor is and how much power to transmit with. Assume that a 25 kHz channel bandwidth is used. This is approximately the bandwidth used by one channel in a GSM system. A 200 kHz wide channel is divided into 8 time slots, which give an effective bandwidth of 25 kHz. Also assume that the receiver has a total noise factor of 10 dB. This gives the receiver noise of

$$N = kTB * NF = -204 \text{ dbW} + 10\log10(25000) + 10$$
$$= -150 \text{ dBW} = -120 \text{ dBm}$$

Fix the transmitter power (Ptx) to 1 W, which is the same as 30 dBm. If the log-normal shadow fading is ignored, the pathgain can be written as,

$$G = -31 - 10\alpha\log10(D)$$

This gives the SNR at a given distance D

$$SNR(D) = Ptx + G - N = 30 - 31 - 10\alpha\log10(D) - (-120)$$

From this it can be seen that if a SNR of at least 10 dB is required, the distance has to be less than approximately 530m.

Each mobile is given a specific power in both the uplink and downlink. To be able to store this data, two more vectors are created: pul for the transmitter power in the uplink, and pdl for the downlink. The mobiles that are not active have their values set to nan as usual. Create the power vectors and assign all active users (they have a channel to communicate on) a power.

```
pul=nans(10,1)
pdl=nans(10,1)
pul(~isnan(k))=30
pdl(~isnan(k))=30
```

There are two functions for calculating the carrier, interference and signal to interference values. They are called *transmitul* and *transmitdl*. One calculates the values for the uplink and the other one for the downlink. The reason that they are separated is that sometimes only the link in one direction is of interest and no calculations have to be wasted on the other direction of the link.

The last input argument is the noise level. The other input arguments are the b and k vectors, the power vector and the gain matrix. Now it should be possible to calculate the values of interest:

```
[cul, iul, sirul]=transmitul(b, k, pul, g, -120)
[cdl, idl, sirdl]=transmitdl(b, k, pdl, g, -120)
```

Since the system created is a fairly small one, there are no co-channel interferers. This means that the interference consists of noise only. This is easily seen in the vectors idl and iul who contain the interference level.

There is a function for plotting cumulative distribution functions: *cdfplotlow*. This can be used to plot the data of interest. Not all mobiles are connected, which means that the C, I, and C/I value cannot be calculated for them. Instead that value is set to nan. While this is a convenient way of handling inactive mobiles, there will be a problem in plotting the data. Therefore there is a small function *clean* that removes all nan values from a vector.

```
cdfplotlow(clean(sirdl))
```

Now it is possible to add the final parts to the simulation program. Insert this code to complete the simulation program.

```
% Assign power to the mobiles
par.pinit=30;
pul=nans(nmob,1);
pdl=nans(nmob,1);
pul(~isnan(k))=par.pinit;
pdl(~isnan(k))=par.pinit;
```

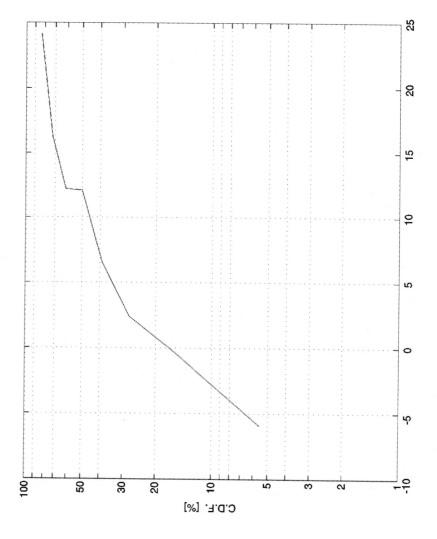

Figure C.3 Example of a cumulative distribution function plot.

```
% Calculate carrier, interference and signal to interference ratio
par.noise=-120;
[cul, iul, sirul]=transmitul(b, k, pul, g, par.noise);
[cdl, idl, sirdl]=transmitdl(b, k, pdl, g, par.noise);



% Plot the cdf of the carrier to interference ratio in the downlink
figure(2);
cdfplotlow(clean(sirdl));
```

About the Authors

Jens Zander is a professor and the head of the Radio Communication Systems Laboratory at the Royal Institute of Technology (KTH) in Stockholm, Sweden. He received an M.Sc. and a Ph.D. in electrical engineering from Linköping University, Linköping, Sweden, in 1979 and 1985, respectively. From 1985 to 1989, he was a partner and in the executive management of SECTRA, a telecommunications company now listed on the Stockholm stock exchange. In 1989 he was appointed to the chair of Radio Communication Systems at the Royal Institute of Technology. Since 1992, Dr. Zander has been a part-time senior scientific advisor to the Swedish Defense Research Agency (FOI). In 2000, he was appointed as the scientific director of the KTH Center for Wireless Systems. Dr. Zander is on the board of several Swedish companies, including TERACOM, the largest Swedish broadcast operator.

Dr. Zander has published about 100 scientific papers and three textbooks in wireless communications. He serves as an associate editor for the ACM journal *Wireless Networks* and as an area editor for *Wireless Personal Communications*. Dr. Zander serves currently as the chairman of the Swedish IEEE VT/COM chapter and is an adjunct member of the Swedish National URSI Committee.

Seong-Lyun Kim is an assistant professor at the Royal Institute of Technology (KTH) in Stockholm, Sweden. He received a B.S. in economics from Seoul National University in 1988, and an M.S. and a Ph.D. in operations research (with applications to radio resource management) from Korea

Advanced Institute of Science and Technology (KAIST) in 1990 and 1994, respectively. His current research and education focus is in radio resource management and economic models for wireless personal communication systems.

Index

Recent Titles in the Artech House Mobile Communications Series

John Walker, Series Editor

Understanding WAP: Wireless Applications, Devices, and Services, Marcel van der Heijden and Marcus Taylor, editors

Universal Wireless Personal Communications, Ramjee Prasad

Wideband CDMA for Third Generation Mobile Communications, Tero Ojanperä and Ramjee Prasad, editors

Wireless Communications in Developing Countries: Cellular and Satellite Systems, Rachael E. Schwartz

Wireless Intelligent Networking, Gerry Christensen, Paul G. Florack, and Robert Duncan

Wireless Technician's Handbook, Andrew Miceli

For further information on these and other Artech House titles, including previously considered out-of-print books now available through our In-Print-Forever® (IPF®) program, contact:

Artech House
685 Canton Street
Norwood, MA 02062
Phone: 781-769-9750
Fax: 781-769-6334
e-mail: artech@artechhouse.com

Artech House
46 Gillingham Street
London SW1V 1AH UK
Phone: +44 (0)20 7596-8750
Fax: +44 (0)20 7630-0166
e-mail: artech-uk@artechhouse.com

Find us on the World Wide Web at:
www.artechhouse.com

CHECK FOR _____1_____ PARTS
(1 CD)